SpringerBriefs in Applied Sciences and Technology

For further volumes:
http://www.springer.com/series/8884

George Jaiani

Cusped Shell-Like Structures

Springer

Prof. Dr. George Jaiani
I. Vekua Institute of Applied Mathematics
Iv. Javakhishvili Tbilisi State University
University Street 2
0186 Tbilisi
Georgia
e-mail: george.jaiani@viam.sci.tsu.ge

ISSN 2191-530X
ISBN 978-3-642-22100-2
e-ISSN 2191-5318
e-ISBN 978-3-642-22101-9
DOI 10.1007/978-3-642-22101-9
Springer Heidelberg Dordrecht London New York

Cover design: eStudio Calamar, Berlin/Figueres

Printed on acid-free paper

Springer is part of Springer Science+Business Media (www.springer.com)

Preface

This work is devoted to an updated exploratory survey of results concerning elastic cusped shells, plates, and beams and cusped prismatic shell-fluid interaction problems. It also contains some up to now non-published results and new problems to be investigated. Mathematically, the corresponding problems lead to non-classical, in general, boundary value and initial-boundary value problems for governing degenerate elliptic and hyperbolic systems in static and dynamical cases, respectively. Two principally different approaches of investigation are used: (1) to get results for 2D (two-dimensional) and 1D (one-dimensional) problems from results of the corresponding 3D (three-dimensional) problems and (2) to investigate directly governing degenerate and singular systems of 2D and 1D problems. In both the cases, it is important to study the relationship of 2D and 1D problems with 3D problems. On the one hand, it turned out that the second approach allows to investigate such 2D and 1D problems whose corresponding 3D problems are not possible to study within the framework of the 3D model of the theory of elasticity. On the other hand, the second approach is historically approved, since first the 1D and 2D models were created and only then the 3D model was constructed. Hence, the second approach gives a good chance for the further development (generalization) of the 3D model.

The present work is addressed to engineers interested in the mathematical aspects of practical problems and mathematicians interested in engineering applications. Both can find new challenging problems expecting their resolution. It will also be very useful for students of advanced courses specializing in mechanics of continua, structural mechanics, mathematical modelling, partial differential equations and applications.

I am greatly indebted to the Scientific Board of EUROMECH 527, and personally to Holm Altenbach for the kind suggestion to write such a work for Springer Briefs Series and to Andreas Oechsner for his support. I express my gratitude to Natalia Chinchaladze and Yusuf Fuat Gülver for their assistance in the preparation of this work and for useful discussions. I acknowledge the support given through travel and research grants for some of the topics surveyed in this work by the NATO Science Programmes, INTAS, CDRF/GRDF, DAAD, DFG, Max-Planck-Society, and the Georgian National Science Foundation.

Last but not the least, I thank my wife Natela for her continuous and unconditional support and understanding.

Tbilisi, Georgia, April 2011 George Jaiani

Contents

Chapter 1
Introduction

The present work is devoted to an updated exploratory survey of investigations concerning elastic *cusped* (in other words, sharpened, cuspidate, cuspate, cuspidal, tapered) *shell-like structures*, namely, cusped (standard and prismatic) shells, plates, and beams, cusped shell-like elastic body–fluid interaction problems. Under cusped shells [see e.g. Vekua (1955, 1985), Jaiani (2001a, 1982), and Chap. 2 below] we understand shells whose thickness vanishes either on a part or on the whole boundary of the standard shell "middle" surface and prismatic shell projection, correspondingly. Beams are called cusped ones [see e.g. Jaiani (2001b, 2002)] if at least at one endpoint of their axes the cross-section area vanishes. Mathematically, we are lead to posing and solving, in the static case, certain boundary value problems (BVPs) for even order equations and systems of elliptic type with order degeneration, while in the dynamical case we face initial boundary value problems (IBVPs) for even order equations and systems of hyperbolic type with order degeneration.

The works surveyed mostly concern the thorough investigation of mathematical problems arised (existence, uniqueness, stability of solutions, in particular cases construction of solutions in the explicit form, etc.) and their application to the original tasks of mechanics (to elastic cusped shell-like structures under consideration). Generally, researches of such a sort we could call mathematical mechanics which together with the technical mechanics (estimate of bounds of applicability of methods and solutions of the strength of materials, problems of stress concentration in zones of sharp variation of shapes of bodies or load distributions, etc.) creates modern mechanics. Elastic cusped shell-like structures as objects of the technical mechanics are less analyzed. It is topical to carry out their research matching needs of engineers; in turn, theoretical importance of the mathematical mechanics is undoubted.

In the study of cusped shell-like bodies, forces concentrated along cusped edges or at cusps may be encountered; moreover, well-posedness of boundary conditions (BCs) in displacements along cusped edges depends on their sharpness geometry.

At present, we have sufficiently complete mathematical theory of elastic cusped prismatic shells and beams; however, the study of cusped standard shells of general form remains topical.

G. Jaiani, *Cusped Shell-Like Structures*, SpringerBriefs in Applied Sciences and Technology, DOI: 10.1007/978-3-642-22101-9_1,

The work is organized as follows.

In Chap. 2 we analyze the geometry of structures under consideration.

Chapter 3 is devoted to the construction of hierarchical models of cusped elastic shells, mainly, prismatic ones, cusped elastic beams, and shallow fluids occupying non-Lipschitz, in general, 3D domains.

Chapter 4 deals with the investigation of hierarchical models of cusped elastic prismatic shells, namely, with explicit solutions for cusped elastic prismatic shell-like bodies with projections of special forms (half-plane, half-strip, infinite plane sector, etc.); variational formulation of the basic 3D problems for elastic prismatic shell-like bodies; approximating function spaces; variational formulation in particular spaces; existence and uniqueness theorems; convergence results; derivation of the basic system of two-dimensional models by means of general systems and the Legendre polynomials (Vekua's system); existence and uniqueness theorems for the governing system of degenerate equations of cusped elastic prismatic shells in the Nth hierarchical model. Cusped Kirchhoff-Love plates are discussed as well.

Chapter 5 deals with the investigation of hierarchical models of cusped beams, namely, with explicit solutions of problems for cusped beams in particular cases; variational formulation of the basic 3D problems for beam-like elastic bodies; approximating function spaces; existence results; convergence results. Moreover, cusped Euler-Bernoulli beams (properties of the general solution of the degenerate Euler-Bernoulli equation; solution of BVPs, harmonic vibration and dynamical problems) are discussed as well.

Chapter 6 is devoted to the study of relations of the hierarchical models of cusped elastic shells and beams to the 3D model.

In Chap. 7 cusped elastic prismatic shell—3D and 2D fluid structure interaction problems are discussed.

The present work is endowed with a wide bibliography connected with topics under consideration.

References

G.V. Jaiani, *Solution of Some Problems for a Degenerate Elliptic Equation of Higher Order and Their Applications to Prismatic Shells* (Tbilisi University Press, Russian, 1982)

G.V. Jaiani, Application of Vekua's dimension reduction method to cusped plates and bars. Bull. TICMI **5**, 27–34 (2001a)

G. Jaiani, On a mathematical model of bars with variable rectangular cross-sections. ZAMM Z. Angew Math. Mech. **81**(3), 147–173 (2001b)

G.V. Jaiani, *Theory of Cusped Euler-Bernoulli Beams and Kirchoff-Love Plates*. Lect. Notes TICMI 3 (2002)

I.N. Vekua, On one method of calculating of prismatic shells (Russian). Trudy Tbilis Mat. Inst. **21**, 191–259 (1955)

I.N. Vekua, On one method of calculating of prismatic shells (Russian). Trudy Tbilis Mat. Inst. **21**, 191–259 (1955)

I.N. Vekua, *Shell Theory: General Methods of Construction* (Pitman Advanced Publishing Program, Boston, 1985)

Chapter 2
Preliminary Topics

Abstract In the present chapter prismatic and cusped prismatic shells are exposed. Relation of the prismatic shells to the standard shells and plates are analyzed. Cusped beams are defined. The Lipschitz boundaries are defined. In a lot of figures 3D illustrations of the cusped prismatic shells and beams are given. Typical cross-sections of cusped prismatic shells and longitudinal sections of the cusped beams are illustrated. Moments of functions and their derivatives are introduced and their relations clarified.

Keywords Cusped prismatic shells · Cusped beams · Moments of functions

2.1 Geometry of Structures Under Consideration

Investigations of cusped elastic prismatic shells actually takes its origin from the fifties of the last century, namely, in 1955, Vekua raised the problem of investigation of elastic cusped prismatic shells, whose thickness on the prismatic shell entire boundary or on its part vanishes [see Vekua (1955) and also Vekua (1985)]. Such bodies, considered as 3D ones, may occupy 3D domains with, in general,

G. Jaiani, *Cusped Shell-Like Structures*, SpringerBriefs in Applied Sciences and Technology, DOI: 10.1007/978-3-642-22101-9_2,

non-Lipschitz boundaries.[1] In practice, such cusped prismatic shells, in particular, cusped plates, and cusped beams (i.e., beams whose cross-sections area vanishes at least at one end of the beam) are often encountered in spatial structures with partly fixed edges, e.g., stadium ceilings, aircraft wings, submarine wings, etc., in machine-tool design, as in cutting-machines, planning-machines, in astronautics, turbines, and in many other application fields of engineering.

Let $Ox_1x_2x_3$ be an anticlockwise-oriented rectangular Cartesian frame of origin O. We conditionally assume the x_3-axis vertical. The elastic body is called a prismatic shell if it is bounded above and below by, respectively, the surfaces (so-called face surfaces)

$$x_3 = \overset{(+)}{h}(x_1, x_2) \text{ and } x_3 = \overset{(-)}{h}(x_1, x_2),$$

laterally by a cylindrical surface Γ of generatrix parallel to the x_3-axis and its vertical dimension is sufficiently small compared with other dimensions of the body.

In other words, the 3D elastic prismatic shell-like body occupies a bounded region $\overline{\Omega}$ with boundary $\partial\Omega$, which is defined as:

$$\Omega := \left\{ (x_1, x_2, x_3) \in \mathbb{R}^3 : (x_1, x_2) \in \omega, \ \overset{(-)}{h}(x_1, x_2) < x_3 < \overset{(+)}{h}(x_1, x_2) \right\}, \tag{2.1}$$

where $\overline{\omega} := \omega \cup \partial\omega$ is the so-called projection of the prismatic shell $\overline{\Omega} := \Omega \cup \partial\Omega$ (see Figs. 2.1, 2.2, 2.3, 2.4, 2.5, 2.6, 2.7 and also Figs. 2.24. 2.25, 2.26, 2.27, 2.28, 2.29, 2.30, 2.31, 2.32, 2.33, 2.34, where typical cross-sections of prismatic shells are given, ν is a normal at O to $\partial\omega$, φ is the angle at the cusp between tangents $\overset{(+)}{T}$ and $\overset{(-)}{T}$, and Figs. 2.8, 2.9, 2.10, 2.11, 2.12); $\gamma := \partial\omega$ and $\partial\Omega$ denote boundaries of ω and Ω, respectively; $\mathbb{R}^n$ is an n-dimensional Euclidian space.

---

[1] The boundary $\partial\Omega$ of an open bounded subset Ω of $\mathbb{R}^n$, $n \geq 2$, is said to be Lipschitz (Lipschitz-continuous) boundary if the following conditions are simultaneously satisfied [see e.g. Ciarlet (1988)]: there exist constants $\alpha > 0$, $\beta > 0$, $\gamma > 0$ and a finite number of local coordinate systems with origins O_r, $r = \overline{1, R}$, coordinates $\zeta'_r := (\xi^r_1, \xi^r_2, \ldots, \xi^r_{n-1})$, and $\zeta_r := \xi^r_n$, and corresponding functions a_r, $r = \overline{1, R}$, such that:

$$\partial\Omega = \bigcup_{r=1}^{R} \{(\zeta'_r, \zeta_r) : \zeta_r = a_r(\zeta'_r), \ |\zeta'_r| < \alpha\};$$

$$\{(\zeta'_r, \zeta_r) : a_r(\zeta'_r) < \zeta_r < a_r(\zeta'_r) + \beta, \ |\zeta'_r| \leq \alpha\} \subset \Omega, \quad r = \overline{1, R};$$

$$\{(\zeta'_r, \zeta_r) : a_r(\zeta'_r) - \beta < \zeta_r < a_r(\zeta'_r), \ |\zeta'_r| \leq \alpha\} \subset \mathbb{R}^n - \overline{\Omega}, \quad r = \overline{1, R};$$

$$|a_r(\zeta'_r) - a_r(\eta'_r)| \leq \gamma |\zeta'_r - \eta'_r| \ \text{ for all } \ |\zeta'_r| \leq \alpha, \ |\eta'_r| \leq \alpha, \ r = \overline{1, R},$$

where the last inequalities express the Lipschitz-continuity of the functions a_r, $r = \overline{1, R}$. The Lipschitz boundary $\partial\Omega$ is necessarily bounded, while this is not necessarily true of the set Ω which can be interchanged with the set $\mathbb{R}^n \backslash \overline{\Omega}$ in the definition. Such a set is called a Lipschitz set.

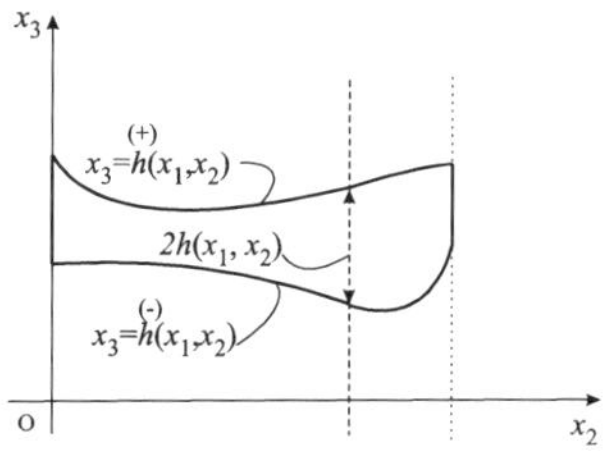

Fig. 2.1 A cross-section of a typical non-cusped prismatic shell having a Lipschitz boundary

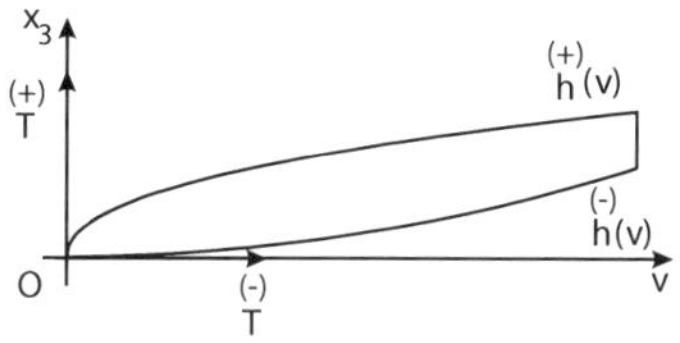

Fig. 2.2 A cross-section of a blunt cusped prismatic shell $(\varphi = \frac{\pi}{2})$. It has a Lipschitz boundary

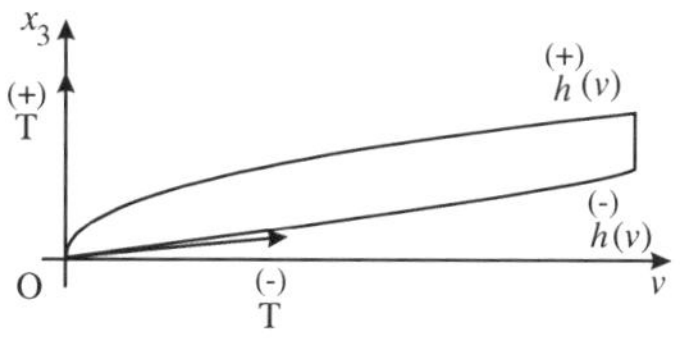

Fig. 2.3 A cross-section of a blunt cusped prismatic shell $(\varphi \in]0, \frac{\pi}{2}[)$. It has a Lipschitz boundary

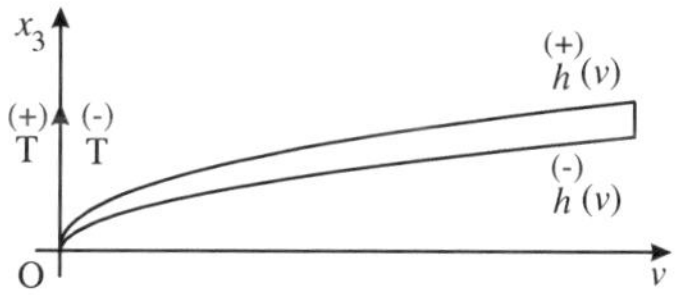

Fig. 2.4 A cross-section of a blunt cusped prismatic shell $(\varphi = 0)$. It has a non-Lipschitz boundary

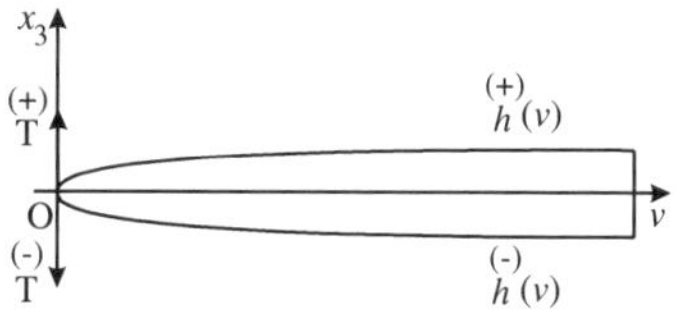

Fig. 2.5 A cross-section of a blunt cusped plate $(\varphi = \pi)$. It has a Lipschitz boundary

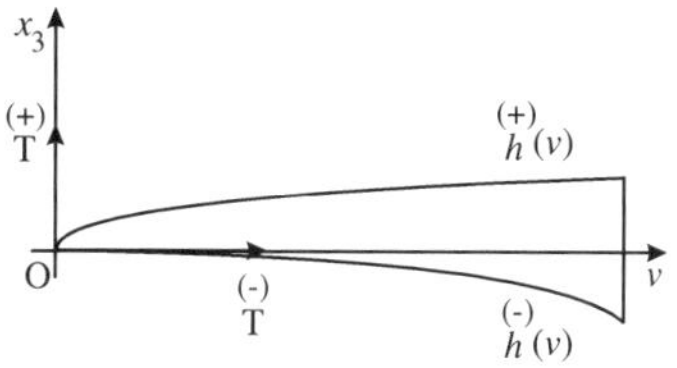

Fig. 2.6 A cross-section of a blunt cusped prismatic shell $(\varphi = \frac{\pi}{2})$. It has a Lipschitz boundary

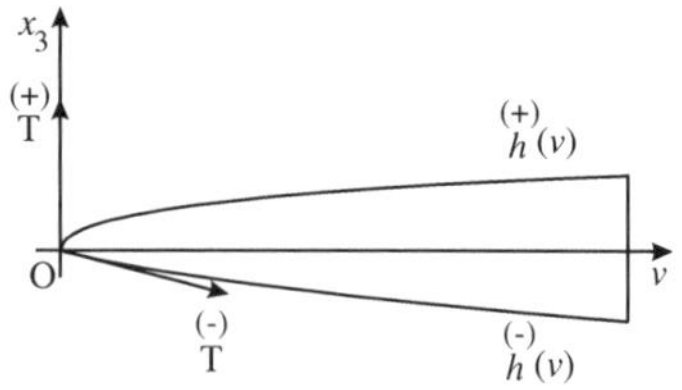

Fig. 2.7 A cross-section of a blunt cusped prismatic shell $(\varphi \in]\frac{\pi}{2}, \pi[)$. It has a Lipschitz boundary

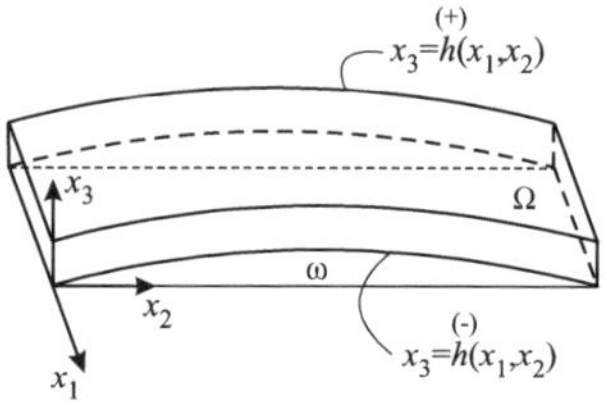

Fig. 2.8 Prismatic shell of a constant thickness. $\partial\Omega$ is a Lipschitz boundary

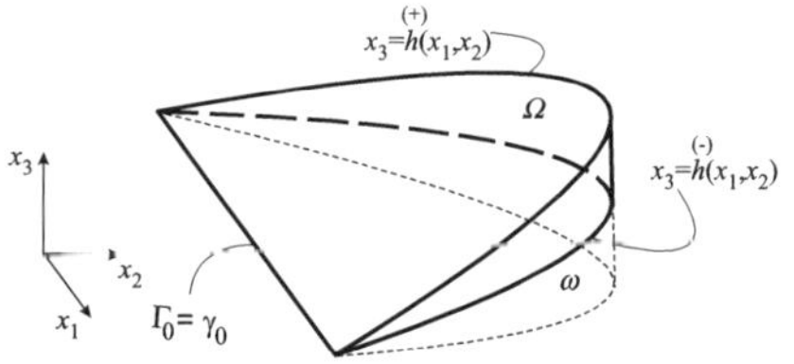

Fig. 2.9 A sharp cusped prismatic shell with a semicircle projection. $\partial\Omega$ is a Lipschitz boundary

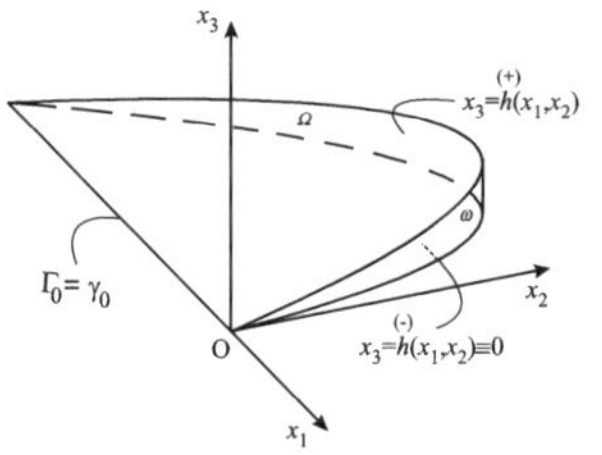

Fig. 2.10 A sharp cusped prismatic shell with a semicircle projection. $\partial\Omega$ is a Lipschitz boundary

In what follows we assume that[2]

$$\overset{(\pm)}{h}(x_1, x_2) \in C^2(\omega) \cap C(\overline{\omega}),$$

[2] $C(\overline{\omega})$ denotes a class of functions continuous on $\overline{\omega}$; $C^2(\omega)$ denotes a class of twice continuously differentiable functions with respect to the variables x_1 and x_2 with $(x_1, x_2) \in \omega$.

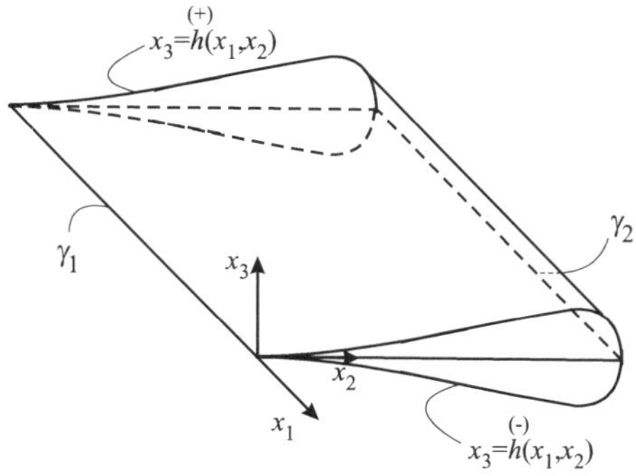

Fig. 2.11 A cusped plate with sharp γ_1 and blunt γ_2 edges, $\gamma_0 = \gamma_1 \cup \gamma_2$. $\partial\Omega$ is a non-Lipschitz boundary

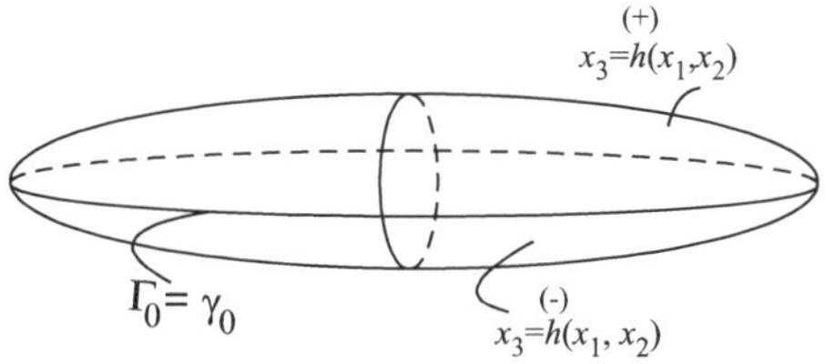

Fig. 2.12 A blunt cusped plate with the edge γ_0. $\partial\Omega$ is a Lipschitz boundary

and

$$2h(x_1,x_2) := \overset{(+)}{h}(x_1,x_2) - \overset{(-)}{h}(x_1,x_2)\begin{cases} > 0 & \text{for} \quad (x_1,x_2) \in \omega, \\ \geq 0 & \text{for} \quad (x_1,x_2) \in \partial\omega \end{cases}$$

is the thickness of the prismatic shell $\overline{\Omega}$ at the points $(x_1,x_2) \in \overline{\omega}$. $\max\{2h\}$ is essentially less than the characteristic dimensions of ω. Let $x_3 = \widetilde{h}(x_1,x_2)$ denote the middle surface of the prismatic shell, then

$$2\widetilde{h}(x_1,x_2) := \overset{(+)}{h}(x_1,x_2) + \overset{(-)}{h}(x_1,x_2).$$

In the symmetric case of the prismatic shells, i.e., when

$$\overset{(-)}{h}(x_1,x_2) = -\overset{(+)}{h}(x_1,x_2), \quad \text{i.e.,} \quad 2\widetilde{h}(x_1,x_2) = 0,$$

we have to do with plates of variable thickness $2h(x_1,x_2)$ and a middle-plane ω (see Figs. 2.11, 2.12). Prismatic shells are called cusped prismatic shells if a set γ_0, consisting of $(x_1,x_2) \in \partial\omega$ for which $2h(x_1,x_2) = 0$, is not empty.

Distinctions between the prismatic shell of a constant thickness and the standard shell of a constant thickness are shown in the Fig. 2.13, where cross-sections

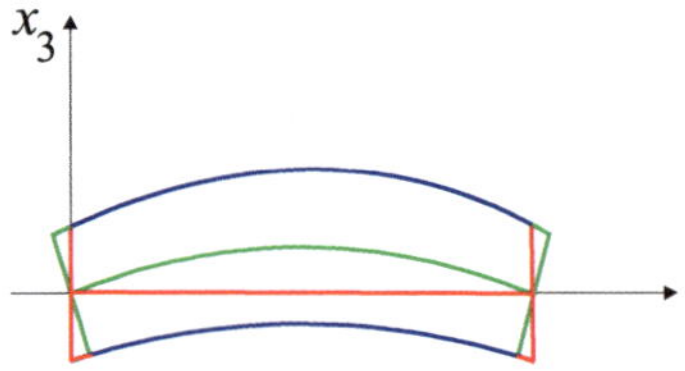

Fig. 2.13 Comparison of cross-sections of prismatic and standard shells

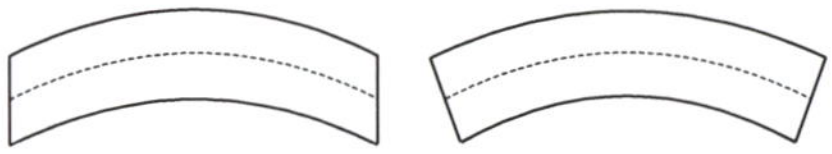

Fig. 2.14 Cross-sections of a prismatic (*left*) and a standard shell with the same mid-surface

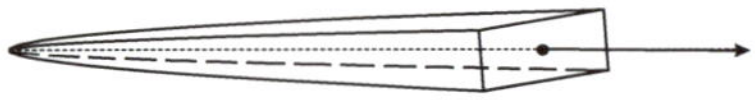

Fig. 2.15 A cusped beam with rectangular cross-sections. $\partial\Omega$ is a Lipschitz boundary

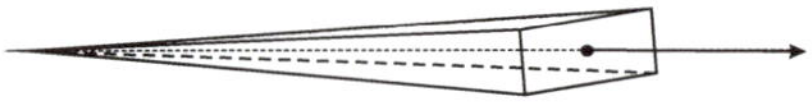

Fig. 2.16 A cusped beam with rectangular cross-sections. $\partial\Omega$ is a non-Lipschitz boundary

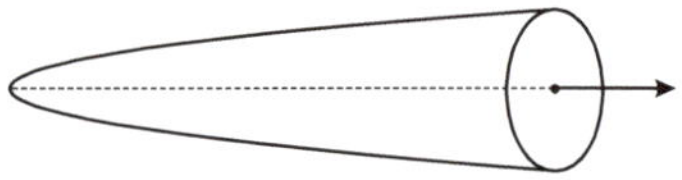

Fig. 2.17 A cusped circular beam. $\partial\Omega$ is a Lipschitz boundary

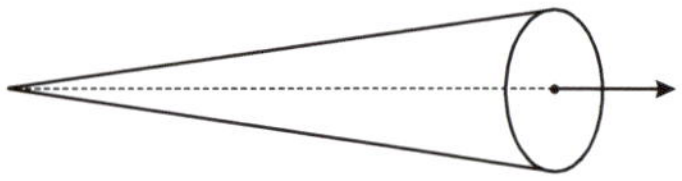

Fig. 2.18 A conical wedge. $\partial\Omega$ is a Lipschitz boundary

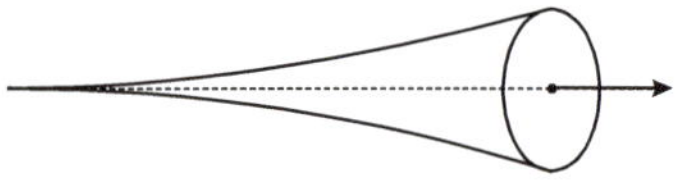

Fig. 2.19 A cusped circular beam. $\partial\Omega$ is a non-Lipschitz boundary

of the prismatic shell of a constant thickness with its projection and of the standard shell of a constant thickness with its middle surface are given in red and green colors, respectively (common parts are given in blue). In other words, the lateral boundary of the standard shell is orthogonal to the "middle surface" of the shell, while the lateral boundary of the prismatic shell is orthogonal to the prismatic shell's projection on $x_3 = 0$ (see also Fig. 2.14).

If in (2.1)

$$\omega = \Big\{ (x_1, x_2) \in \mathbb{R}^2 : 0 < x_1 < L, \overset{(-)}{h}_2(x_1) < x_2 < \overset{(+)}{h}_2(x_1),$$
$$0 \le 2h_2(x_1) := \overset{(+)}{h}_2(x_1) - \overset{(-)}{h}_2(x_1) << L, L = const > 0 \Big\},$$

then the prismatic shell-like body $\overline{\Omega}$ will become a beam-like body with a cross-section of an arbitrary form. Such a beam will be called a cusped beam if the cross-section area of the beam equals zero at least at one of its end [see Figs. 2.15, 2.16, 2.17, 2.18, 2.19, 2.20, 2.21, 2.22, 2.23 and Figs. 2.24, 2.25, 2.26, 2.27, 2.28, 2.29, 2.30, 2.31, 2.32, 2.33, 2.34, where typical longitudinal (vertical and horizontal) sections of beams are given]. In particular, let a domain $\overline{\Omega}^b$ of $\mathbb{R}^3$ occupied by an elastic beam be

$$\Omega^b := \Big\{ (x_1, x_2, x_3) \in \mathbb{R}^3 : 0 < x_1 < L, \overset{(-)}{h}_i(x_1) < x_i < \overset{(+)}{h}_i(x_1),$$
$$2h_i(x_1) := \overset{(+)}{h}_i - \overset{(-)}{h}_i \ge 0, h_i \in C([0,L]) \bigcap C^2(]0,L[), i=2,3, L=const > 0 \Big\}$$

and $2h_3$ and $2h_2$ be correspondingly the thickness and the width of the beam and their maxima be essentially less then the length L of the bar; superscript "b" means beam. If at least one of the conditions $2h_i(0)=0$ and $2h_i(L)=0$, $i=2,3$, is fulfilled, a beam is called the cusped beam. The last class of beams consists of beams with rectangular cross-sections which may degenerate in segments or points (see Figs. 2.15, 2.16, 2.20, 2.21, 2.22, 2.23) at the beams ends.

2.2 Moments of Functions

Let $f(x_1, x_2, x_3)$ be a given function in $\overline{\Omega}$ having integrable partial derivatives, f_r be its rth order moment defined as follows

$$f_r(x_1, x_2) := \int_{\overset{(-)}{h}(x_1, x_2)}^{\overset{(+)}{h}(x_1, x_2)} f(x_1, x_2, x_3) P_r(ax_3 - b) dx_3,$$

where

$$a(x_1, x_2) := \frac{1}{h(x_1, x_2)}, \quad b(x_1, x_2) := \frac{\widetilde{h}(x_1, x_2)}{h(x_1, x_2)},$$

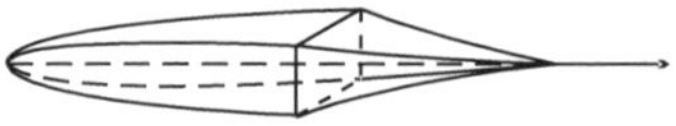

Fig. 2.20 A cusped beam with rectangular cross-sections. $\partial\Omega$ is a non-Lipschitz boundary

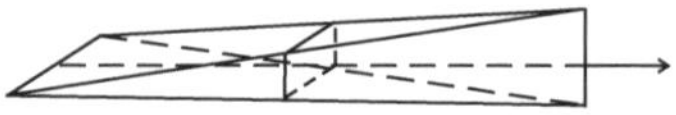

Fig. 2.21 A cusped beam with rectangular cross-sections. $\partial\Omega$ is a Lipschitz boundary

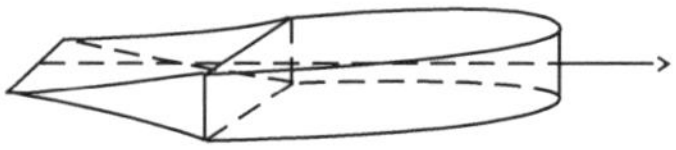

Fig. 2.22 A cusped beam with rectangular cross-sections. $\partial\Omega$ is a non-Lipschitz boundary

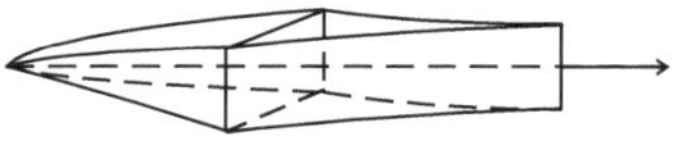

Fig. 2.23 A cusped beam with rectangular cross-sections. $\partial\Omega$ is a non-Lipschitz boundary

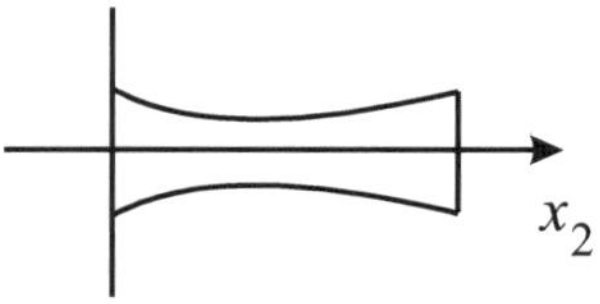

Fig. 2.24 Non-cusped ends (*edges*)

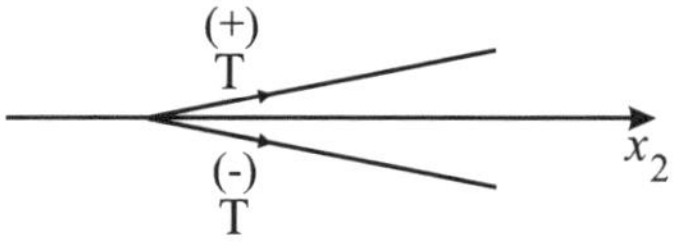

Fig. 2.25 Wedge, $\varphi \in]0, \pi[$

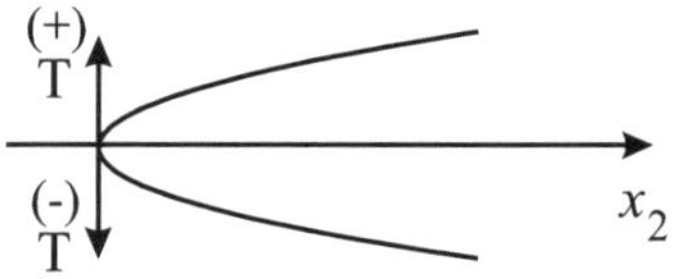

Fig. 2.26 $\varphi = \pi$

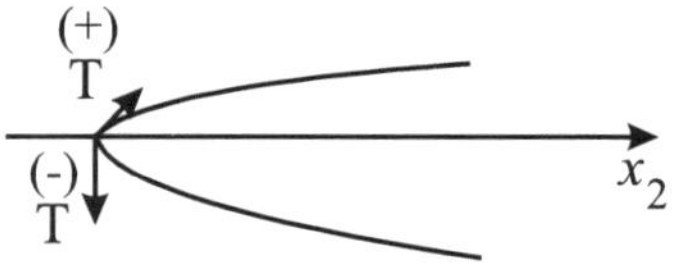

Fig. 2.27 $\frac{\pi}{2} < \varphi < \pi$

Fig. 2.28 $\frac{\pi}{2} < \varphi < \pi$

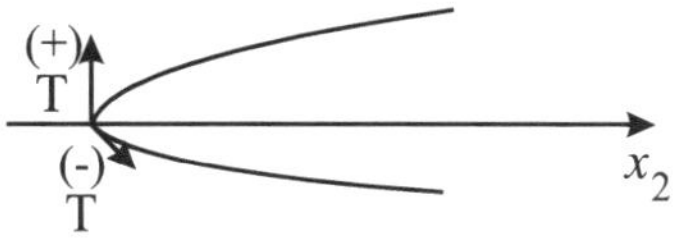

Fig. 2.29 $\varphi = \frac{\pi}{2}$

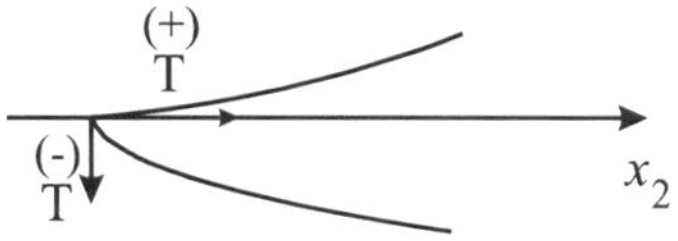

Fig. 2.30 $\varphi = \frac{\pi}{2}$

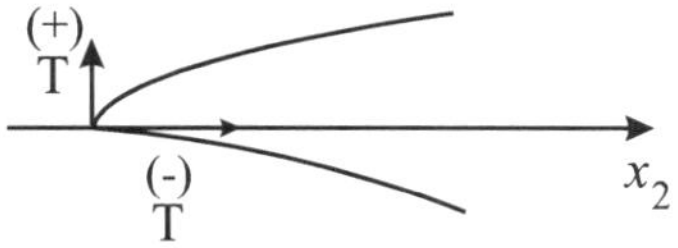

Fig. 2.31 $0 < \varphi < \frac{\pi}{2}$

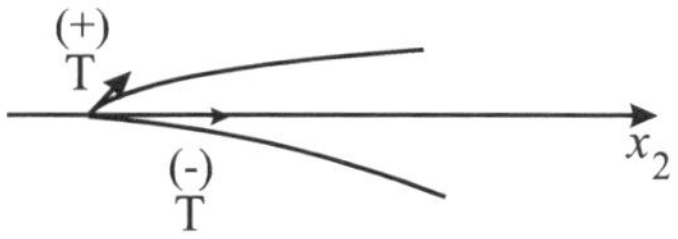

Fig. 2.32 $0 < \varphi < \frac{\pi}{2}$

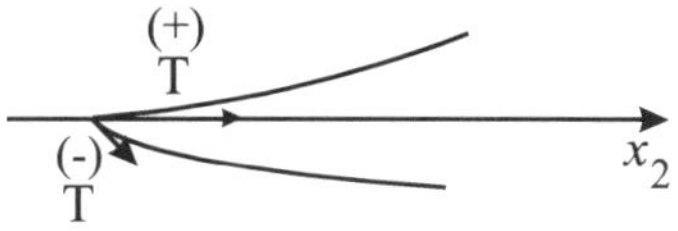

Fig. 2.33 $0 < \varphi < \pi$

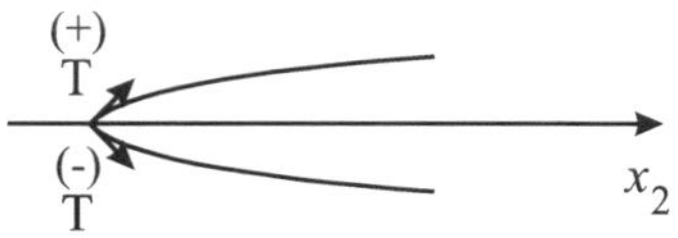

Fig. 2.34 $\varphi = 0$

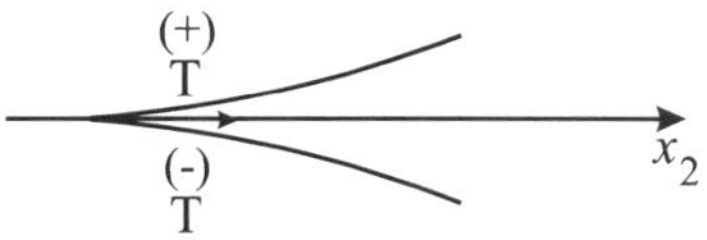

and

$$P_r(\tau) = \frac{1}{2^r r!}\frac{d^r(\tau^2-1)^r}{d\tau^r}, \quad r = 0, 1, \ldots,$$

are the rth order Legendre Polynomials with the orhogonality property

$$\int\limits_{-1}^{+1} P_m(\tau)P_n(\tau)d\tau = \frac{2}{2m+1}\delta_{mn}.$$

Using the well-known formulas of integration by parts (with respect to x_3) and differentiation with respect to a parameter of integrals depending on parameters (x_α), taking into account $P_r(1) = 1$, $P_r(-1) = (-1)^r$, we deduce

$$\int\limits_{\overset{(-)}{h}(x_1,x_2)}^{\overset{(+)}{h}(x_1,x_2)} P_r(ax_3-b)f_{,3}\,dx_3 = -a\int\limits_{\overset{(-)}{h}(x_1,x_2)}^{\overset{(+)}{h}(x_1,x_2)} P'_r(ax_3-b)fdx_3 + \overset{(+)}{f} - (-1)^r\overset{(-)}{f}, \tag{2.2}$$

$$\int\limits_{\overset{(-)}{h}(x_1,x_2)}^{\overset{(+)}{h}(x_1,x_2)} P_r(ax_3-b)f_{,\alpha}\,dx_3 = f_{r,\alpha} - \overset{(+)}{f}\overset{(+)}{h}_{,\alpha} + (-1)^r\overset{(-)}{f}\overset{(-)}{h}_{,\alpha}$$

$$-\int\limits_{\overset{(-)}{h}(x_1,x_2)}^{\overset{(+)}{h}(x_1,x_2)} P'_r(ax_3-b)(a_{,\alpha}x_3-b_{,\alpha})fdx_3, \quad \alpha = 1, 2, \tag{2.3}$$

where superscript prime means differentiation with respect to the argument $ax_3 - b$, subscripts preceded by a comma mean partial derivatives with respect to the corresponding variables, $\overset{(\pm)}{f} := f[x_1, x_2, \overset{(\pm)}{h}(x_1,x_2)]$. Applying the following relations from the theory of the Legendre polynomials [see e.g. Jaiani (2004), p. 299]

$$\begin{aligned} P'_r(\tau) &= \sum_{s=0}^{r}(2s+1)\frac{1-(-1)^{r+s}}{2}P_s(\tau), \\ \tau P'_r(\tau) &= rP_r(\tau) + P'_{r-1}(\tau) = rP_r(\tau) + \sum_{s=0}^{r-1}(2s+1)\frac{1+(-1)^{r+s}}{2}P_s(\tau) \end{aligned} \tag{2.4}$$

and, in view of $\frac{a_{,\alpha}}{a} = -\frac{h_{,\alpha}}{h}$, $\frac{a_{,\alpha}}{a}b = \widetilde{h}a_{,\alpha}$, $b_{,\alpha} = (\widetilde{h}a)_{,\alpha}$, it is easily seen that

$$P'_r(ax_3 - b)(a_{,\alpha} x_3 - b_{,\alpha}) = \frac{a_{,\alpha}}{a}(ax_3 - b)P'_r(ax_3 - b) + (\frac{a_{,\alpha}}{a} b - b_{,\alpha})P'_r(ax_3 - b)$$
$$= -h_{,\alpha} h^{-1}(ax_3 - b)P'_r(ax_3 - b) - \widetilde{h}_{,\alpha} h^{-1} P'_r(ax_3 - b)$$
$$= -\overset{r}{a}_{\alpha \underline{r}} P_r(ax_3 - b) - \sum_{s=0}^{r-1} \overset{r}{a}_{\alpha s} P_s(ax_3 - b), \qquad (2.5)$$

where

$$\overset{r}{a}_{\alpha \underline{r}} := r\frac{h_{,\alpha}}{h}, \quad \overset{r}{a}_{\alpha s} := (2s+1)\frac{\overset{(+)}{h}_{,\alpha} - (-1)^{r+s} \overset{(-)}{h}_{,\alpha}}{2h}, \quad s \neq r.$$

Now, bearing in mind (2.5) and (2.4), from (2.2) and (2.3) we have

$$\int_{\overset{(-)}{h}(x_1,x_2)}^{\overset{(+)}{h}(x_1,x_2)} P_r(ax_3 - b) f_{,\alpha}\, dx_3 = f_{r,\alpha} + \sum_{s=0}^{r} \overset{r}{a}_{\alpha s} f_s - \overset{(+)}{f}\,\overset{(+)}{h}_{,\alpha} + (-1)^r \overset{(-)}{f}\,\overset{(-)}{h}_{,\alpha}, \quad \alpha = 1, 2, \qquad (2.6)$$

$$\int_{\overset{(-)}{h}(x_1,x_2)}^{\overset{(+)}{h}(x_1,x_2)} P_r(ax_3 - b) f_{,3}\, dx_3 = \sum_{s=0}^{r} \overset{r}{a}_{3s} f_s + \overset{(+)}{f} - (-1)^r \overset{(-)}{f}, \qquad (2.7)$$

respectively. Here

$$\overset{r}{a}_{3s} := -(2s+1)\frac{1 - (-1)^{s+r}}{2h}.$$

References

P.G. Ciarlet, *Mathematical Elasticity, Vol I: Three-dimensional elasticity* (North-Holland Amsterdam, 1988)

G. Jaiani, *Mathematical Models of Mechanics of Continua* (in Georgian) (Tbilisi University Press, Tbilisi, 2004)

I.N. Vekua, On one method of calculating of prismatic shells (Russian). Trudy Tbilis Mat. Inst. **21**, 191–259 (1955)

I.N. Vekua, *Shell Theory: General Methods of Construction* (Pitman Advanced Publishing Program, Boston, 1985)

Chapter 3
Hierarchical Models

Abstract The present chapter contains hierarchical models for elastic prismatic shells, beams with rectangular variable cross-sections, fluids and elastic solid–fluid structures occupying prismatic domains. To this end, a dimension reduction method based on Fourier–Legendre expansions is mostly applied to basic equations of linear theory of elasticity of homogeneous isotropic bodies and Newtonian fluids. The governing equations and systems of hierarchical models are constructed with respect to so-called mathematical moments of stress and strain tensors and displacement vector components.

Keywords Hierarchical models · Shells · Plates · Beams · Mathematical moments · Elasticity · Newtonian fluid

3.1 Vekua's Hierarchical Models of the First Type for Elastic Prismatic Shells

In what follows X_{ij} and e_{ij} are the stress and strain tensors, respectively, u_i are the displacements, Φ_i are the volume force components, ρ is the density, λ and μ are the Lamé constants, δ_{ij} is the Kronecker delta. Moreover, repeated indices imply summation (Greek letters run from 1 to 2, and Latin letters run from 1 to 3, unless otherwise stated), bar under one of the repeated indices means that we do not sum.

By u_{ir}, X_{ijr}, e_{ijr}, Φ_{jr} we denote the rth order moments of the corresponding quantities u_i, X_{ij}, e_{ij}, Φ_j as defined below:

G. Jaiani, *Cusped Shell-Like Structures*, SpringerBriefs in Applied Sciences and Technology, DOI: 10.1007/978-3-642-22101-9_3,

$$\left(u_{ir}, X_{ijr}, e_{ijr}, \Phi_{jr}\right)(x_1, x_2, t)$$

$$:= \int\limits_{\overset{(-)}{h}(x_1,x_2)}^{\overset{(+)}{h}(x_1,x_2)} \left(u_i, X_{ij}, e_{ij}, \Phi_j\right)(x_1, x_2, x_3, t) P_r(ax_3 - b) dx_3, \quad i,j = \overline{1,3}. \quad (3.1)$$

Vekua's hierarchical models for elastic prismatic shells are the mathematical models (Vekua 1955, 1985). Their constructing is based on the multiplication of the basic equations of linear elasticity:

Motion Equations

$$X_{ij,i} + \Phi_j = \rho \ddot{u}_j(x_1, x_2, x_3, t), \quad (x_1, x_2, x_3) \in \Omega \subset \mathbb{R}^3, \quad t > t_0, \quad j = \overline{1,3}; \quad (3.2)$$

Generalized Hooke's law (isotropic case)

$$X_{ij} = \lambda \theta \delta_{ij} + 2\mu e_{ij}, \quad i,j = \overline{1,3}, \quad \theta := e_{ii}; \quad (3.3)$$

Kinematic Relations

$$e_{ij} = \frac{1}{2}(u_{i,j} + u_{j,i}), \quad i,j = \overline{1,3}, \quad (3.4)$$

by Legendre polynomials $P_r(ax_3 - b)$ and then integration with respect to x_3 within the limits $\overset{(-)}{h}(x_1, x_2)$ and $\overset{(+)}{h}(x_1, x_2)$. By constructing Vekua's hierarchical models in Vekua's first version on upper and lower face surfaces stress-vectors are assumed to be known, while there the values of the displacements

$$\overset{(\pm)}{u}_i := u_i(x_1, x_2, \overset{(\pm)}{h}(x_1, x_2), t) = \sum_{s=0}^{\infty} a\left(s + \frac{1}{2}\right) u_{is} (\pm 1)^s = \sum_{s=0}^{\infty} \frac{(\pm 1)^s (2s+1)}{2h} u_{is}$$

are calculated from their (displacements') Fourier–Legendre series

$$\left(u_i, X_{ij}, e_{ij}\right)(x_1, x_2, x_3, t) := \sum_{r=0}^{\infty} a\left(r + \frac{1}{2}\right) \left(u_{ir}, X_{ijr}, e_{ijr}\right)(x_1, x_2, t) P_r(ax_3 - b) \quad (3.5)$$

expansions on the segment $x_3 \in \left[\overset{(-)}{h}(x_1, x_2), \overset{(+)}{h}(x_1, x_2)\right]$ [this will be used by deriving (3.8); clearly,

$$\overset{(+)}{u}_i - (-1)^r \overset{(-)}{u}_i = -\sum_{s=0}^{r} a_{3s} u_{is}, \quad \overset{(+)}{u}_i \overset{(+)}{h}_{,\alpha} - (-1)^r \overset{(-)}{u}_i \overset{(-)}{h}_{,\alpha} = \sum_{s=0}^{r} a^*_{\alpha s} u_{is}, \quad i = \overline{1,3},$$

where $\overset{r}{a}{}^{*}_{\alpha s} = \overset{r}{a}_{\alpha s}, \quad s \neq r, \quad \overset{r}{a}{}^{*}_{\alpha r} = (2r+1)\frac{h_{,\alpha}}{h}\Big]$ and vice versa in his second version. The volume force components Φ_i and, therefore, their rth order moments are assumed to be known in both the versions. Using preliminary derived formulas (2.6), (2.7), from (3.2) to (3.4), we get, respectively,

$$X_{\alpha jr,\alpha} + \sum_{s=0}^{r} \overset{r}{a}_{is} X_{ijs} + \overset{r}{X}_j = \rho \frac{\partial^2 u_{jr}}{\partial t^2}, \quad j = \overline{1,3}, \quad r = 0,1,\ldots, \tag{3.6}$$

$$X_{ijr}(x_1,x_2,t) = \lambda \delta_{ij} \theta_r(x_1,x_2,t) + 2\mu e_{ijr}(x_1,x_2,t), \quad i,j = \overline{1,3}, \quad r = 0,1,\ldots, \tag{3.7}$$

$$e_{ijr} = \frac{1}{2}\left(u_{ir,j} + u_{jr,i}\right) + \frac{1}{2}\sum_{s=r}^{\infty} \overset{r}{b}_{is} u_{js} + \frac{1}{2}\sum_{s=r}^{\infty} \overset{r}{b}_{js} u_{is}, \quad i,j = \overline{1,3}, \quad r = 0,1,\ldots, \tag{3.8}$$

where

$$\theta_r := e_{iir} = u_{\gamma r,\gamma} + \sum_{s=r}^{\infty} \overset{r}{b}_{is} u_{is}, \quad \overset{r}{b}_{\alpha \underline{r}} := -(r+1)\frac{h_{,\alpha}}{h}, \quad \overset{r}{b}_{3\underline{r}} = 0,$$

$$\overset{r}{b}_{js} := \begin{cases} 0, & s<r, \\ -\overset{r}{a}_{\alpha s} = -(2s+1)\dfrac{\overset{(+)}{h}_{,\alpha} - (-1)^{r+\underline{s}}\overset{(-)}{h}_{,\alpha}}{2h}, & j=\alpha, \quad s>r, \\ (2s+1)\frac{1-(-1)^{\underline{s}+r}}{2h}, & j=3, \quad s>r, \end{cases}$$
$$\alpha = 1,2, \quad j = \overline{1,3}, \quad r,s = 0,1,2,\ldots;$$

$$\overset{r}{X}_j := \overset{(+)}{X}_{3j} - \overset{(+)}{X}_{\alpha j}\overset{(+)}{h}_{,\alpha} + (-1)^r\left[-\overset{(-)}{X}_{3j} + \overset{(-)}{X}_{\alpha j}\overset{(-)}{h}_{,\alpha}\right] + \Phi_{jr} = Q_{\overset{(+)}{n}j}\sqrt{1+\left(\overset{(+)}{h}_{,1}\right)^2 + \left(\overset{(+)}{h}_{,2}\right)^2}$$
$$+ (-1)^r Q_{\overset{(-)}{n}j}\sqrt{1+\left(\overset{(-)}{h}_{,1}\right)^2 + \left(\overset{(-)}{h}_{,2}\right)^2} + \Phi_{jr}, \quad j = \overline{1,3}, \quad r = 0,1,2,\ldots;$$

$Q_{\overset{(+)}{n}j}$ and $Q_{\overset{(-)}{n}j}$ are components of the stress vectors acting on the upper and lower face surfaces with normals $\overset{(+)}{n}$ and $\overset{(-)}{n}$, respectively. So, we get the equivalent[1] to (3.2)–(3.4) infinite system (3.6)–(3.8) with respect to the so-called rth order moments X_{ijr}, e_{ijr}, u_{ir}. Then, substituting (3.8) in (3.7) and the obtained in (3.6), we construct an equivalent infinite system with respect to the rth order moments u_{ir} (Vekua 1955). After this, if we suppose that the moments whose subscripts, indicating moments' order, are greater than N equal zero and consider only the first $N+1$ equations ($r = \overline{0,N}$) in the obtained infinite system of equations with respect to the rth order moments u_{ir}, we obtain the Nth order approximation (hierarchical model) governing system consisting of $3N+3$ equations with respect to $3N+3$ unknown functions $\overset{N}{u}_{ir}$ (roughly speaking $\overset{N}{u}_{ir}$ is an "approximate value" of u_{ir}, since $\overset{N}{u}_{ir}$ are solutions of the derived finite system), $i = \overline{1,3}, r = \overline{0,N}$. Because of $\overset{r}{b}_{3\underline{r}} = 0, h^{r+1}(h^{-r-1})_{,\alpha} = \overset{r}{b}_{\alpha\underline{r}}, \alpha = 1,2,$ we can rewrite (3.8) for $v_{ir} := h^{-r-1}u_{ir}$ as follows

$$e_{ijr} = \frac{1}{2}h^{r+1}\left(v_{ir,j} + v_{jr,i}\right) + \frac{1}{2}\sum_{s=r+1}^{\infty} h^{s+1}\left(\overset{r}{b}_{is}v_{js} + \overset{r}{b}_{js}v_{is}\right), \quad i,j = \overline{1,3}, \ r = 0,1,\ldots. \tag{3.9}$$

In particular,

$$\theta_r := e_{iir} = h^{r+1}v_{\gamma r,\gamma} + \sum_{s=r+1}^{\infty} h^{s+1}\overset{r}{b}_{is}v_{is}, \qquad r = 0,1,\ldots.$$

Multiplying equality (3.6) by h^r and, taking into account that $\overset{r}{a}_{i\underline{r}} = rh^{-1}h_{,\alpha}$, we get

$$(h^r X_{\alpha jr})_{,\alpha} + h^r \sum_{s=0}^{r-1} \overset{r}{a}_{is} X_{ijs} + h^r \overset{r}{X}_j = \rho h^r \frac{\partial^2 h^{r+1} v_{jr}}{\partial t^2}, \quad j = \overline{1,3}, \quad r = 0,1,\ldots. \tag{3.10}$$

Substituting (3.9) in (3.7) and then the obtained

$$X_{ijr} = \lambda\delta_{ij}h^{r+1}v_{\gamma r,\gamma} + \mu h^{r+1}(v_{ir,j} + v_{jr,i}) + \sum_{s=r+1}^{\infty} \overset{r}{B}_{ijks}h^{s+1}v_{ks}, \ i,j = \overline{1,3}, \quad r = 0,1,\ldots,$$

---

[1] In the following sense: if X_{ij}, e_{ij}, and u_i satisfy the relations (3.2)–(3.4), then constructed by (3.1) functions X_{ijr}, e_{ijr}, u_{ir} will satisfy the infinite relations (3.6)–(3.8) and, vice versa, if X_{ijr}, e_{ijr}, u_{ir} satisfy the infinite relations (3.6)–(3.8), then constructed by means of (3.5) functions X_{ij}, e_{ij}, u_i will satisfy the relations (3.2)–(3.4).

where

$$\overset{r}{B}_{ijks} := \lambda\delta_{ij}\overset{r}{b}_{ks} + \mu\delta_{kj}\overset{r}{b}_{is} + \mu\delta_{ik}\overset{r}{b}_{js},$$

in (3.10), we derive the infinite system which in the Nth approximation of the first version of Vekua's hierarchical models for cusped, in general, homogeneous elastic prismatic shells reads as (see Jaiani 2001a; Vekua 1965):

$$\begin{aligned}
&\mu\left[\left(h^{2r+1}\overset{N}{v}_{\alpha r,j}\right)_{,\alpha} + \left(h^{2r+1}\overset{N}{v}_{jr,\alpha}\right)_{,\alpha}\right] + \lambda\delta_{\alpha j}\left(h^{2r+1}\overset{N}{v}_{\gamma r,\gamma}\right)_{,\alpha}\\
&+ \sum_{s=r+1}^{N}\left(\overset{r}{B}_{\alpha jks}\, h^{r+s+1}\overset{N}{v}_{ks}\right)_{,\alpha} + \sum_{l=0}^{r-1}\overset{r}{a}_{il}\left[\lambda\delta_{ij}h^{r+l+1}\overset{N}{v}_{\gamma l,\gamma} + \mu h^{r+l+1}\left(\overset{N}{v}_{il,j} + \overset{N}{v}_{jl,i}\right)\right.\\
&\left.+ \sum_{s=l+1}^{N}\overset{l}{B}_{ijks}\, h^{r+s+1}\overset{N}{v}_{ks}\right] + h^{r}\overset{r}{X}_{j} = \rho h^{r}\frac{\partial^2 h^{r+1}\overset{N}{v}_{jr}}{\partial t^2}, \quad r=\overline{0,N},\ j=\overline{1,3},\ \sum_{q}^{q-1}(\ldots)\equiv 0,
\end{aligned} \tag{3.11}$$

where

$$\overset{N}{v}_{kr} := \frac{\overset{N}{u}_{kr}}{h^{r+1}}, \quad k=\overline{1,3}, \quad r=\overline{0,N} \tag{3.12}$$

(in what follows we omit superscripts like N when it does not lead to misunderstanding) are unknown so-called weighted "moments" of displacements. Having $\overset{N}{u}_{kr}$, we can calculate $\overset{N}{e}_{ijr}$ and $\overset{N}{X}_{ijr}$ by means of (3.8), (3.7).

In the Nth approximation for the approximate values of 3D displacements u_i, stress X_{ij} and strain e_{ij} tensors we take the expressions (3.5), where in the sum r runs from 0 to N and u_{ir}, X_{ijr}, and e_{ijr} are replaced by $\overset{N}{u}_{ir}$, $\overset{N}{X}_{ijr}$, and $\overset{N}{e}_{ijr}$ $(i,j=\overline{1,3}, r=\overline{0,N})$ correspondingly.

If λ and μ depend on $(x_1,x_2)\in\omega$, the principal part of (3.11) will have the following form:

$$(\mu h^{2r+1}\overset{N}{v}_{\alpha r,j})_{,\alpha} + (\mu h^{2r+1}\overset{N}{v}_{jr,\alpha})_{,\alpha} + \delta_{\alpha j}(\lambda h^{2r+1}\overset{N}{v}_{\gamma r,\gamma})_{,\alpha}, \quad j=\overline{1,3}, \quad r=\overline{0,N}.$$

The remained part of (3.11) can be rewritten in the appropriate form.

Hierarchical models for standard shells of variable thickness one can find in (Vekua 1985, 1965).

3.2 Vekua's Hierarchical Models of the Second Type for Elastic Prismatic Shells

Similar hierarchical models, when on the face surfaces displacements are prescribed can be constructed analogously. In the Nth approximation the corresponding governing system has the following form

$$
\begin{aligned}
&\Big[\mu(u_{\alpha r,\beta}+u_{\beta r,\alpha})\Big]_{,\alpha}+(\lambda u_{\gamma r,\gamma})_{,\beta}+\Big(\lambda\sum_{s=0}^{r}\overset{r}{a}_{ks}u_{ks}\Big)_{,\beta}\\
&+\Big[\mu\sum_{s=0}^{r}(\overset{r}{a}_{\alpha s}u_{\beta s}+\overset{r}{a}_{\beta s}u_{\alpha s})\Big]_{,\alpha}+\sum_{s=r}^{N}\overset{r}{b}_{is}\Big[\lambda\delta_{i\beta}u_{\gamma s,\gamma}+\mu(u_{is,\beta}+u_{\beta s,i})\Big]\\
&+\sum_{s=r}^{N}\overset{r}{b}_{is}\sum_{s'=0}^{s}\Big[\lambda\delta_{i\beta}\overset{s}{a}_{ks'}u_{ks'}+\mu(\overset{s}{a}_{is'}u_{\beta s'}+\overset{s}{a}_{\beta s'}u_{is'})\Big]+\delta_{\alpha\beta}(\lambda\overset{r}{\Psi}_{kk})_{,\alpha}+2(\mu\overset{r}{\Psi}_{\alpha\beta})_{,\alpha}\\
&+\sum_{s=r}^{N}\overset{r}{b}_{is}(\lambda\delta_{i\beta}\overset{s}{\Psi}_{kk}+2\mu\overset{s}{\Psi}_{i\beta})+\Phi_{\beta r}=\rho\frac{\partial^2 u_{\beta r}}{\partial t^2},\quad \beta=1,2,\quad r=\overline{0,N};\\
&(\mu u_{3r,\alpha})_{,\alpha}+\Big[\mu\sum_{s=0}^{r}(\overset{r}{a}_{\alpha s}u_{3s}+\overset{r}{a}_{3s}u_{\alpha s})\Big]_{,\alpha}\\
&+\sum_{s=r}^{N}\overset{r}{b}_{is}(\lambda\delta_{i3}u_{\gamma s,\gamma}+\mu u_{3s,i})+\sum_{s=r}^{N}\overset{r}{b}_{is}\sum_{s'=0}^{s}\Big[\lambda\delta_{i3}\overset{s}{a}_{ks'}u_{ks'}+\mu(\overset{s}{a}_{is'}u_{3s'}+\overset{s}{a}_{3s'}u_{is'})\Big]\\
&+2(\mu\overset{r}{\Psi}_{\alpha 3})_{,\alpha}+\sum_{s=r}^{N}\overset{r}{b}_{is}(\lambda\delta_{i3}\overset{r}{\Psi}_{kk}+2\mu\overset{r}{\Psi}_{i3})+\Phi_{3r}=\rho\frac{\partial^2 u_{3r}}{\partial t^2},\quad r=\overline{0,N},
\end{aligned}
\tag{3.13}
$$

where

$$
\begin{aligned}
\overset{r}{\Psi}_{\alpha\beta}&:=-\frac{1}{2}\Big[\overset{(+)}{u}_{\alpha}\overset{(+)}{h}_{,\beta}-(-1)^r\overset{(-)}{u}_{\alpha}\overset{(-)}{h}_{,\beta}\Big]-\frac{1}{2}\Big[\overset{(+)}{u}_{\beta}\overset{(+)}{h}_{,\alpha}-(-1)^r\overset{(-)}{u}_{\beta}\overset{(-)}{h}_{,\alpha}\Big],\\
&\qquad\alpha,\beta=1,2;\\
\overset{r}{\Psi}_{\alpha 3}=\overset{r}{\Psi}_{3\alpha}&:=\frac{1}{2}\Big[\overset{(+)}{u}_{\alpha}-(-1)^r\overset{(-)}{u}_{\alpha}\Big]-\frac{1}{2}\Big[\overset{(+)}{u}_{3}\overset{(+)}{h}_{,\alpha}-(-1)^r\overset{(-)}{u}_{3}\overset{(-)}{h}_{,\alpha}\Big],\quad\alpha=1,2;\\
\overset{r}{\Psi}_{33}&:=\overset{(+)}{u}_{3}-(-1)^r\overset{(-)}{u}_{3}.
\end{aligned}
$$

In particular, for $N=0$ and weighted zero moments v_{j0}, $j=\overline{1,3}$, we have

$$
\begin{aligned}
&\mu\big[(hv_{\alpha 0})_{,\beta}+(hv_{\beta 0})_{,\alpha}\big]_{,\beta}+\lambda\big[(hv_{\gamma 0})_{,\gamma}\big]_{,\alpha}-(\ln h)_{,\beta}\Big\{\lambda\delta_{\alpha\beta}(hv_{\gamma 0})_{,\gamma}\\
&+\mu\big[(hv_{\alpha 0})_{,\beta}+(hv_{\beta 0})_{,\alpha}\big]\Big\}+2\mu\ \Psi_{\alpha\beta,\beta}(x_1,x_2,t)+\lambda\Psi_{kk,\alpha}(x_1,x_2,t)-(\ln h)_{,\beta}\\
&\times\big[\lambda\delta_{\alpha\beta}\Psi_{kk}(x_1,x_2,t)+2\mu\ \Psi_{\alpha\beta}(x_1,x_2,t)\big]+\Phi_{\alpha 0}(x_1,x_2,t)=\rho h\frac{\partial^2 v_{\alpha 0}}{\partial t^2},\quad\alpha=1,2;\\
&\ \mu(hv_{30})_{,\beta\beta}-(\ln h)_{,\beta}\,\mu(hv_{30})_{,\beta}+2\mu\ \Psi_{3\beta,\beta}(x_1,x_2,t)\\
&\ \ -2\mu(\ln h)_{,\beta}\,\Psi_{3\beta}(x_1,x_2,t)+\Phi_{30}(x_1,x_2,t)=\rho h\frac{\partial^2 v_{30}}{\partial t^2},
\end{aligned}
\tag{3.14}
$$

where

$$\Psi_{33}(x_1,x_2,t) := u_3(x_1,x_2,\overset{(+)}{h},t) - u_3(x_1,x_2,\overset{(-)}{h},t),$$

$$2\Psi_{i\beta}(x_1,x_2,t) = u_i\left(x_1,x_2,\overset{(-)}{h},t\right)\overset{(-)}{h}_{,\beta} - u_i\left(x_1,x_2,\overset{(+)}{h},t\right)\overset{(+)}{h}_{,\beta}$$

$$+\begin{cases} -u_\beta\left(x_1,x_2,\overset{(+)}{h},t\right)\overset{(+)}{h}_{,\alpha} + u_\beta\left(x_1,x_2,\overset{(-)}{h},t\right)\overset{(-)}{h}_{,\alpha} & \text{for } i=\alpha, \quad \alpha=1,2; \\ u_\beta\left(x_1,x_2,\overset{(+)}{h},t\right) - u_\beta\left(x_1,x_2,\overset{(-)}{h},t\right) & \text{for } i=3. \end{cases}$$

Remark 3.1 In the same manner we can construct hierarchical models when either on one face surface the surface forces, while on the another one the displacements are prescribed or neither the surface forces nor the displacements are prescribed on one or both the face surfaces. The last hierarchical models one can find in Vekua (1955).

3.3 Hierarchical Models for Elastic Beams With Variable Rectangular Cross-Sections

The analogues system in the $(N_3,\ N_2)$ approximation of hierarchical models for cusped, in general, beams with variable rectangular cross-sections are derived by Jaiani (2001b):

$$\Lambda_{\underline{j}}\left(h_2^{\underline{2n_2+1}}\, h_3^{\underline{2n_3+1}}\overset{(N_3,N_2)}{v}_{jn_3n_2,1}\right)_{,1}$$
$$+\sum_{i=1}^{3}\sum_{r=0}^{N_3}\sum_{s=0}^{N_2}\left(R^{ij}_{rs}\overset{(N_3,N_2)}{v}_{irs,\,1} + S^{ij}_{rs}\overset{(N_3,N_2)}{v}_{irs}\right) + h_2^{\underline{n_2}}\, h_3^{\underline{n_3}}\overset{n_3,n_2}{X^0_j} \tag{3.15}$$
$$= \rho h_2^{\underline{n_2}} h_3^{\underline{n_3}}\frac{\partial^2 h_2^{\underline{n_2+1}}\, h_3^{\underline{n_3+1}}\overset{(N_3,N_2)}{v}_{jn_3n_2}}{\partial t^2}, \quad j=\overline{1,3}; \quad n_k=\overline{0,N_k}, \quad k=2,3,$$

where $\overset{(N_3,N_2)}{v}_{jn_3n_2}$ are unknown so-called weighted double "moments" of displacements [see (3.17) below], $h_2(x_1)h_3(x_1) \geq 0$ for $x_1=0$, $x_1=L$ (i.e., the beam may be cusped one),

$$\Lambda_j := \begin{cases} \lambda + 2\mu, & j=1, \\ \mu, & j=2,3, \end{cases}$$

$R^{ij}_{rs}(x_1), S^{ij}_{rs}(x_1)$ are expressed by means of λ, μ, $\overset{(+)}{h}_i(x_1)$, $\overset{(-)}{h}_i(x_1)$, $i=2,3$; some of $R^{ij}_{rs}(x_1)$, $S^{ij}_{rs}(x_1)$ can be zero and some of them cannot be bounded on]0,L[,

$\overset{n_3,n_2}{X_j^0}$ are expressed by external forces acting on the face surfaces $x_i = \overset{(\pm)}{h_i}(x_1)$, $i = 2, 3$, and double moments of volume forces.

3.4 Relation of Mathematical Moments and Fourier Coefficients

If $N = +\infty$, then, by virtue of (3.1), keeping in mind (3.12)

$$v_{kl}(x_1,x_2) = \overset{\infty}{v}_{kl}(x_1,x_2) := h^{-l-1}(x_1,x_2)\overset{\infty}{u}_{kl}(x_1,x_2) = h^{-l-1}(x_1,x_2)u_{kl}(x_1,x_2)$$

$$:= h^{-l-1}(x_1,x_2) \int\limits_{\overset{(-)}{h}(x_1,x_2)}^{\overset{(+)}{h}(x_1,x_2)} u_k(x_1,x_2,x_3,t)P_l(a(x_1,x_2)x_3 - b(x_1,x_2))dx_3, \quad l = 1,2,\ldots, \tag{3.16}$$

where $u_k, k = \overline{1,3}$, are 3D displacements.

If $N_i = +\infty, i = 2, 3$, then the double moments (see Jaiani 2001b)

$$\begin{aligned} v_{kn_3n_2}(x_1) &= \overset{(\infty,\infty)}{v}_{kn_3n_2}(x_1) := h_2^{-n_2-1}(x_1)h_3^{-n_3-1}(x_1)\overset{(\infty,\infty)}{u}_{kn_3n_2}(x_1) \\ &= h_2^{-n_2-1}(x_1)h_3^{-n_3-1}(x_1)u_{kn_3n_2}(x_1) \\ &= h_2^{-n_2-1}(x_1)h_3^{-n_3-1}(x_1) \int\limits_{\overset{(-)}{h_2}(x_1)}^{\overset{(+)}{h_2}(x_1)} \int\limits_{\overset{(-)}{h_3}(x_1)}^{\overset{(+)}{h_3}(x_1)} u_k(x_1,x_2,x_3,t) \\ &\quad \times P_{n_2}(a_2(x_1)x_2 - b_2(x_1))P_{n_3}(a_3(x_1)x_3 - b_3(x_1))dx_3dx_2, \end{aligned} \tag{3.17}$$

where $a_i(x_1) := \frac{1}{h_i(x_1)}, \quad b_i(x_1) := \frac{\widetilde{h}_i(x_1)}{h_i(x_1)}$,

$2h_i(x_1) := \overset{(+)}{h_i}(x_1) - \overset{(-)}{h_i}(x_1), 2\widetilde{h}_i(x_1) := \overset{(+)}{h_i}(x_1) + \overset{(-)}{h_i}(x_1), i = 2, 3.$

For sufficiently smooth $u_k(x_1,x_2,x_3,t)$ [e.g., $u_k \in C^2(\overline{\Omega})]^2$,

$k = \overline{1,3}$,

[2] We remind that $C^2(\Omega)$ denotes a class of functions twice continuously differentiable with respect to the variables x_1, x_2, x_3, $(x_1,x_2,x_3) \in \Omega$. Note that for the uniform convergence of the above Fourier–Legendre expansions it suffices to demand this property only with respect to $x_3 \in \left[\overset{(-)}{h}(x_1,x_2), \overset{(+)}{h}(x_1,x_2)\right]$ and $x_i \in [\overset{(-)}{h_i}(x_1), \overset{(+)}{h_i}(x_1)], i = 2, 3$, respectively.

$$u_k(x_1,x_2,x_3,t)=\sum_{r=0}^{\infty}\left(k+\frac{1}{2}\right)h^{-1}(x_1,x_2)u_{kr}(x_1,x_2,t)P_r(a(x_1,x_2)x_3-b(x_1,x_2)),$$

[Note that the Fourier coefficient (with respect to the orthonormal system of functions $(k+1/2)^{1/2}h^{-1/2}P_r(a(x_1,x_2)x_3-b(x_1,x_2))$

$$\overset{F}{u}_{kr}(x_1,x_2,t)=\left(k+\frac{1}{2}\right)^{\frac{1}{2}}h^{-\frac{1}{2}}(x_1,x_2)u_{kr}(x_1,x_2,t)\Big],$$

$$u_k(x_1,x_2,x_3,t)=\sum_{n_2,n_3=0}^{\infty}\left(n_2+\frac{1}{2}\right)\left(n_3+\frac{1}{2}\right)h_2^{-1}(x_1)h_3^{-1}(x_1)u_{kn_3n_2}(x_1,t)$$
$$\times P_{n_2}(a_2(x_1)x_2-b_2(x_1))P_{n_3}(a_3(x_1)x_3-b_3(x_1)),\quad k=\overline{1,3},$$

correspondingly to the prismatic shell and beam.

For the finite approximations N and (N_3,N_2) under $\overset{N}{v}_{kr},\overset{N}{u}_{kr}$, and $\overset{(N_2,N_3)}{v}_{kn_3n_2}$ we understand solutions of systems (3.11), (3.13), and (3.15), correspondingly. They can be represented as (3.16) and (3.17), respectively, only if $u_k(x_1,x_2,x_3)$ is a polynomial with respect to x_3 (of the order $\leq N$) and x_2,x_3 (correspondingly of the order $\leq N_2$ and $\leq N_3$), respectively. In other words, for finite N (excluding the case when u_k are polynomials with respect to x_3 of the order $\leq N$) $\overset{N}{v}_{kl}$, $k=\overline{1,3},l=\overline{0,N}$, denote solutions of the system (3.11) and the last equality in (3.16) for $\overset{N}{u}_{kl}$ is not true. Nevertheless, prescribed in BCs $\overset{N}{v}_{kl}$, $k=\overline{1,3},l=\overline{0,N}$, are calculated by (3.16) provided that u_k, $k=\overline{1,3}$, are prescribed on the lateral boundary of the prismatic shell. The same should be stated concerning $\overset{(N_3,N_2)}{v}_{kn_3n_2}$.

3.5 Bi-Modular Prismatic Rods

Governing system in the (N_3,N_2) approximation of the bi-modular prismatic rods has the following form (compare with (3.15) and see (Jaiani 2001b)):

$$\left(h_2^{2n_2+1}h_3^{2n_3+1}\Lambda_j\left(\frac{v_{100,1}(x_1,t)}{|v_{100,1}(x_1,t)|}\right)v_{jn_3n_2,1}\right)_{,1}+\sum_{i=1}^{3}\sum_{r=0}^{N_3}\sum_{s=0}^{N_2}(R_{rs}^{ij}\,v_{irs,\,1}+S_{rs}^{ij}\,v_{irs})$$
$$+h_2^{n_2}h_3^{n_3}\overset{n_3,n_2}{X_j^0}=\rho h_2^{n_2}h_3^{n_3}\frac{\partial^2 h_2^{n_2+1}h_3^{n_3+1}v_{jn_3n_2}}{\partial t^2},$$
$$0<x_1<L,\quad j=\overline{1,3};\quad n_i=\overline{0,N_i},\quad i=2,3,\tag{3.18}$$

where

$$\Lambda_j\left(\frac{v_{100,1}(x_1,t)}{|v_{100,1}(x_1,t)|}\right):=\begin{cases}\lambda+2\mu=\frac{E(1-\nu)}{(1+\nu)(1-2\nu)}=\frac{E\left(\frac{v_{100,1}(x_1,t)}{|v_{100,1}(x_1,t)|}\right)(1-\nu)}{(1+\nu)(1-2\nu)}, & j=1,\\ \mu=\frac{E\left(\frac{v_{100,1}(x_1,t)}{|v_{100,1}(x_1,t)|}\right)}{2(1+\nu)}, & j=2,3,\end{cases}$$

$$E\left(\frac{v_{100,1}(x_1,t)}{|v_{100,1}(x_1,t)|}\right)=\begin{cases}E=\text{const when } v_{100,1}(x_1,t)<0,\\ \beta^2E, \beta\in[0,1[, E=\text{const when } v_{100,1}(x_1,t)>0.\end{cases}$$

If $\beta\neq 1$, system (3.18) is nonlinear. If in addition $h_j(x_1)\geq 0$, $j=2,3$, for $x_1=0$, $x_1=L$, then the rod may be cusped one. Consequently, system (3.18) may be degenerate one. Such a material corresponds to a no-tension material if $\beta=0$ and to a linear elastic one if $\beta=1$. Let us note that in our considerations Hooke's law for rod-like 3D bodies looks like:

$$X_{ij}(x_1,x_2,x_3,t)=\frac{E\left(\frac{v_{100,1}(x_1,t)}{|v_{100,1}(x_1,t)|}\right)\nu}{(1+\nu)(1-2\nu)}\left[e_{11}(x_1,t)+\sum_{k=2}^{3}e_{kk}(x_1,x_2,x_3,t)\right]\delta_{ij}$$
$$+\frac{E\left(\frac{v_{100,1}(x_1,t)}{|v_{100,1}(x_1,t)|}\right)}{(1+\nu)}e_{ij}(x_1,x_2,x_3,t),\quad i,j=\overline{1,3}.$$

In the (0,0) approximation (3.18) takes the following form

$$\left(h_2h_3\Lambda_j\left(\frac{v_{1,1}(x_1,t)}{|v_{1,1}(x_1,t)|}\right)v_{j,1}(x_1,t)\right)_{,1}+\overset{0,0}{X_j^0}=\rho h_2h_3\frac{\partial^2 v_j(x_1,t)}{\partial t^2},\quad j=\overline{1,3},$$

where

$$v_j(x_1,t):=v_{j00}(x_1,t)=\frac{u_{j00}(x_1,t)}{h_2(x_1)h_3(x_1)},\quad j=\overline{1,3}.$$

If

$$v_{j00}(x_1,t)\equiv 0,\ \ h_j=const,\ \ j=2,3;\ \ \overset{0,0}{X_j^0}(x_1,t)\equiv 0,\ \ j=\overline{1,3},$$

and Poisson's ratio $\nu=0$, we get the well-known governing equation of longitudinal oscillations of bi-modular rods (see Lucchesi 2005)

$$\frac{\partial^2 v_1(x_1,t)}{\partial t^2}-\tilde{k}^2(\zeta)\frac{\partial^2 v_1(x_1,t)}{\partial x_1^2}=0,$$

where

$$\zeta := \frac{v_{1,1}(x_1,t)}{|v_{1,1}(x_1,t)|}, \quad \tilde{k}(\zeta) = \begin{cases} k \; if \; \zeta = -1, \\ \beta k \; if \; \zeta = 1, \end{cases} \quad k := \sqrt{\frac{E}{\rho}}, \quad \beta \in [0,1[.$$

System (3.18) is not yet investigated.

3.6 Hierarchical Models of the First Type for Fluids

Now, applying the above Vekua's dimension reduction method, we construct hierarchical models for shallow fluids occupying non-Lipschitz, in general, prismatic domains within the scheme of small displacements linearized with respect to the rest state (see Chinchaladze and Jaiani 2007). As it is well known, motion of the Newtonian fluid is characterized by the following equations

$$\rho^f \ddot{u}_i^f(x_1,x_2,x_3,t) = \sigma_{ij,j}^f(x_1,x_2,x_3,t) + \Phi_i^f(x_1,x_2,x_3,t), \quad i = \overline{1,3}, \tag{3.19}$$

$$\sigma_{ij}^f = -\delta_{ij}p + \lambda^f \delta_{ij}\dot{\theta}^f(\dot{u}^f) + 2\mu^f \dot{e}_{ij}^f(\dot{u}^f), \quad i,j = \overline{1,3}, \tag{3.20}$$

$$\dot{e}_{ij}^f := \frac{1}{2}\left(\dot{u}_{i,j}^f + \dot{u}_{j,i}^f\right), \quad i,j = \overline{1,3}, \tag{3.21}$$

$$\dot{\theta}^f := \dot{e}_{ii}^f = \dot{u}_{i,i}^f = \mathrm{div}\dot{u}^f, \tag{3.22}$$

where $u^f := \left(u_1^f, u_2^f, u_3^f\right)$ is a displacement vector, σ_{ij}^f is a stress tensor, e_{ij}^f is a strain tensor, p is a pressure, Φ_i^f, $i = \overline{1,3}$, are components of the volume force, λ^f and μ^f are the coefficients of viscosity, ρ^f is a density of the fluid, superscript f means fluid, and superscript dot means differentiation with respect to t.

In the case of incompressible barotropic fluids, to the system (3.19)–(3.21) we add the continuity equation

$$\mathrm{div}\dot{u}^f = 0, \tag{3.23}$$

and the state equation

$$\chi(\rho^f, p) = 0,$$

where χ is a certain function defining the state equation.

Let the fluid occupy prismatic domain and the upper and lower face surfaces of the prismatic domain be given by $x_3 = \overset{(+)}{h^f}(x_1,x_2)$ and $x_3 = \overset{(-)}{h^f}(x_1,x_2)$, respectively. Let further

$$2h^f(x_1,x_2) := \overset{(+)}{h^f}(x_1,x_2) - \overset{(-)}{h^f}(x_1,x_2)$$

denote the thickness of the domain occupied by the fluid,

$$2\widetilde{h}^f(x_1,x_2) := \overset{(+)}{h^f}(x_1,x_2) + \overset{(-)}{h^f}(x_1,x_2)$$

and stress vector components on the upper and lower face surfaces be assumed to be known.

Multiplying equations (3.19)–(3.22) by $P_r(a^f x_3 - b^f)$, where

$$a^f(x_1,x_2) := \frac{1}{h^f(x_1,x_2)}, \quad b^f(x_1,x_2) := \frac{\widetilde{h}^f(x_1,x_2)}{h^f(x_1,x_2)},$$

and integrating with respect to x_3 within the limits $\overset{(-)}{h^f}$ and $\overset{(+)}{h^f}$, for the rth order moments with respect to the Legendre polynomials we get

$$\overset{r}{\sigma^f_{\alpha j r,\alpha}} + \sum_{s=0}^{r} \overset{r}{a^f_{is}} \sigma^f_{jis} + \overset{r}{X^f_j} = \rho \ddot{u}^f_{jr}, \tag{3.24}$$

$$\sigma^f_{ijr}(x_1,x_2,t) = -\delta_{ij} p_r(x_1,x_2,t) + \lambda^f \delta_{ij} \dot{\theta}^f_r(x_1,x_2,t) + 2\mu^f \dot{e}^f_{ijr}(x_1,x_2,t), \tag{3.25}$$

$$\dot{e}^f_{ijr} = \frac{1}{2}\sum_{s=r}^{\infty} \overset{r}{b^f_{is}} \dot{u}^f_{js} + \frac{1}{2}\sum_{s=r}^{\infty} \overset{r}{b^f_{js}} \dot{u}^f_{is} + \overset{\dot{r}}{E^f_{ij}}, \quad i,j=\overline{1,3}, \quad r=0,1,\ldots, \tag{3.26}$$

where

$$\dot{\theta}^f_r := \dot{e}^f_{iir} = \dot{u}^f_{\gamma r,\gamma} + \sum_{s=r}^{\infty} \overset{r}{b^f_{is}} \dot{u}^f_{is}, \tag{3.27}$$

$$\overset{r}{b^f_{\alpha \underline{r}}} := -(r+1)\frac{h^f_{,\alpha}}{h^f}, \quad \overset{r}{b^f_{3\underline{r}}} = 0, \quad \overset{r}{b^f_{js}} := \begin{cases} 0, & s<r, \\ -\overset{r}{a^f_{js}}, & s>r, \end{cases} \tag{3.28}$$

$$\begin{aligned}
&\overset{\dot{r}}{E^f_{ij}} := \frac{1}{2}\left(\dot{u}^f_{ir,j} + \dot{u}^f_{jr,i}\right), \quad \overset{r}{a^f_{\alpha s}} := (2s+1)\frac{\overset{(+)}{h^f_{,\alpha}} - (-1)^{r+s}\overset{(-)}{h^f_{,\alpha}}}{2h^f}, \quad s \neq r; \\
&\overset{r}{a^f_{\alpha \underline{r}}} := r\frac{\overset{(+)}{h^f_{,\alpha}} - \overset{(-)}{h^f_{,\alpha}}}{2h^f} = r\frac{h^f_{,\alpha}}{h^f}, \quad \overset{r}{a^f_{3s}} := -(2s+1)\frac{1-(-1)^{r+s}}{2h^f}, \quad \alpha=1,2, \\
&\overset{r}{X^f_j} := \overset{(+)}{\sigma^f_{3j}} - \overset{(+)}{\sigma^f_{\gamma j}}\overset{(+)}{h^f_{,\gamma}} + (-1)^r\left[-\overset{(-)}{\sigma^f_{3j}} + \overset{(-)}{\sigma^f_{\gamma j}}\overset{(-)}{h^f_{,\gamma}}\right] + \Phi^f_{jr} = Q^f_{\overset{(+)}{n}j}\sqrt{1+\left(\overset{(+)}{h^f_{,1}}\right)^2 + \left(\overset{(+)}{h^f_{,2}}\right)^2} \\
&\quad + (-1)^r Q^f_{\overset{(-)}{n}j}\sqrt{1+\left(\overset{(-)}{h^f_{,1}}\right)^2 + \left(\overset{(-)}{h^f_{,2}}\right)^2} + \Phi^f_{jr}, \quad j=\overline{1,3}, \quad r=0,1,\ldots,
\end{aligned} \tag{3.29}$$

$\underset{n\ j}{Q^f_{(+)}}$ and $\underset{n\ j}{Q^f_{(-)}}$ are components of the stress vectors acting on the upper and lower face surfaces. By Φ^f_{jr} we denote the rth moments of the components of the volume forces.

After subsituting first (3.25) into (3.24) and then (3.26), (3.27) into the obtained, we get

$$
\begin{aligned}
&-\delta_{\alpha j}p_{r,\alpha} + \lambda^f \delta_{\alpha j}\dot{\theta}^f_{r,\alpha} + 2\mu^f \dot{e}_{\alpha jr,\alpha} + \sum_{s=o}^{r} \overset{r}{d}{}^f_{is}\left(-\delta_{ij}p_s + \lambda^f \delta_{ij}\dot{\theta}^f_s + 2\mu^f \dot{e}_{ijs}\right)\\
&+ \overset{r}{X}{}^f_i = \rho^f \ddot{u}^f_{jr}, \quad j=\overline{1,3}, \quad r=0,1,\ldots,
\end{aligned}
$$

and

$$
\begin{aligned}
&\lambda^f \delta_{\alpha j}\dot{u}^f_{\beta r,\beta\alpha} + \lambda^f \delta_{\alpha j}\left(\sum_{s=r}^{\infty} \overset{r}{b}{}^f_{is}\,\dot{u}^f_{is}\right)_{,\alpha} + \mu^f\left(\sum_{s=r}^{\infty} \overset{r}{b}{}^f_{\alpha s}\,\dot{u}^f_{js}\right)_{,\alpha} + \mu^f\left(\sum_{s=r}^{\infty} \overset{r}{b}{}^f_{js}\,\dot{u}^f_{\alpha s}\right)_{,\alpha}\\
&+ 2\mu^f \overset{\dot r}{E}{}^f_{\alpha j,\alpha} + \sum_{s=o}^{r} \overset{r}{d}{}^f_{is}\left(\lambda^f \delta_{ij}\left[\dot{u}^f_{\alpha s,\alpha} + \sum_{s_1=s}^{\infty} \overset{s}{b}{}^f_{ks_1}\,\dot{u}^f_{ks_1}\right] + \mu^f\left[\sum_{s_1=s}^{\infty} \overset{s}{b}{}^f_{is_1}\,\dot{u}^f_{js_1}\right.\right.\\
&\left.\left. + \sum_{s_1=s}^{\infty} \overset{s}{b}{}^f_{js_1}\,\dot{u}^f_{is_1} + \dot{u}^f_{js,i} + \dot{u}^f_{is,j}\right]\right) + \overset{r}{X}{}^f_j = \rho^f \ddot{u}^f_{jr} + \delta_{\alpha j}p_{r,\alpha} + \delta_{ij}\sum_{s=0}^{r} \overset{r}{d}{}^f_{is}\,p_s, \quad j=\overline{1,3},
\end{aligned}
$$

respectively. Whence, we obtain the following system

$$
\begin{aligned}
&\mu^f \Delta\dot{u}^f_{\alpha r} + \left(\lambda^f + \mu^f\right)\overset{\dot r}{E}{}^f_{,\alpha} + \overset{r}{M}{}^f_{\alpha} + \overset{r}{X}{}^f_{\alpha} = \rho^f \ddot{u}^f_{\alpha r} + p_{r,\alpha} + \sum_{s=0}^{r} \overset{r}{d}{}^f_{\alpha s}\,p_s, \qquad \alpha = 1,2,\\
&\mu^f \Delta\dot{u}^f_{3r} + \overset{r}{M}{}^f_3 + \overset{r}{X}{}^f_3 = \rho^f \ddot{u}^f_{3r} + \sum_{s=0}^{r} \overset{r}{d}{}^f_{3s}\,p_s,
\end{aligned}
$$

where

$$
\begin{aligned}
&\overset{\dot r}{E}{}^f := \overset{\dot r}{E}{}^f_{ii} = \dot{u}^f_{\alpha r,\alpha},\\
&\overset{r}{M}{}^f_j := \left(\sum_{s=r}^{N} \overset{r}{B}{}^f_{\alpha jks}\,\dot{u}^f_{ks}\right)_{,\alpha} + \sum_{s=0}^{r}\left(\lambda^f\, \overset{r}{d}{}^f_{js}\,\overset{\dot s}{E}{}^f + 2\mu^f\, \overset{r}{d}{}^f_{is}\,\overset{\dot s}{E}{}^f_{ij} + \sum_{q=0}^{N} \overset{r}{d}{}^f_{is}\,\overset{s}{B}{}^f_{ijkq}\,\dot{u}^f_{kq}\right), \; j=\overline{1,3},\\
&\overset{r}{B}{}^f_{ijks} := \lambda^f \delta_{ij}\,\overset{r}{b}{}^f_{ks} + \mu^f \delta_{kj}\,\overset{r}{b}{}^f_{is} + \mu^f \delta_{ik}\,\overset{r}{b}{}^f_{js}, \quad i,j,k,s=\overline{1,3}, \quad r=0,1,\ldots.
\end{aligned}
$$

From (3.23), taking into account (3.22) and (3.27), it follows

$$
\dot{u}^f_{\gamma r,\gamma} + \sum_{s=r}^{\infty} \overset{r}{b}{}^f_{is}\,\dot{u}^f_{is} = 0, \quad r=0,1,2,\ldots.
$$

If we restrict ourself to values $r = \overline{0, N}$, then for $N = 0, 1, \ldots$, we get, so-called, hierarchical differential models corresponding to the system (3.19)–(3.21), (3.23).

3.7 Hierarchical Models of the Second Type for Fluids

Now, we derive hierarchical models when on the face surfaces velocities are assumed to be known. Using the following expansion

$$\sigma^f_{ij} = \sum_{r=0}^{\infty} a^f \left(r + \frac{1}{2} \right) \sigma^f_{ijr} P_r(a^f x_3 - b^f), \quad i,j = \overline{1,3},$$

for $j = \overline{1,3}$, we obtain

$$\overset{(+)}{\sigma^f_{3j}} - (-1)^r \overset{(-)}{\sigma^f_{3j}} - \overset{(+)}{\sigma^f_{\alpha j}} \overset{(+)}{h^f_{,\alpha}} + (-1)^r \overset{(-)}{\sigma^f_{\alpha j}} \overset{(-)}{h^f_{,\alpha}} = - \sum_{s=0}^{\infty} \overset{r}{a^{*f}_{is}} \sigma^f_{ijs}, \tag{3.30}$$

where $\overset{r}{a^{*f}_{\alpha s}} = \overset{r}{a^f_{\alpha s}}$, when $s \neq r$; $\overset{r}{a^{*f}_{\alpha \underline{r}}} = (2r+1) \frac{h^f_{,\alpha}}{h}$, $\alpha = 1, 2$, $s, r = 0, 1, \ldots$.

Substituting (3.30) into (3.29), from (3.24) we get

$$\sigma^f_{\alpha jr,\alpha} + \sum_{s=r}^{\infty} \overset{r}{b^f_{is}} \sigma^f_{ijs} + \Phi^f_{jr} = \rho^f \ddot{u}_{jr}, \quad j = \overline{1,3}, \quad r = 0, 1, \ldots, \tag{3.31}$$

where $\overset{r}{b^f_{is}}$ are given by the formulas (3.28). Using these formulas, (3.31) can be rewritten as follows

$$h^{r+1} \left(h^{-r-1} \sigma^f_{\alpha jr} \right)_{,\alpha} + \sum_{s=r+1}^{\infty} \overset{r}{b^f_{is}} \sigma^f_{ijs} + \Phi^f_{jr} = \rho^f \ddot{u}_{jr}, \quad j = \overline{1,3}, \quad r = 0, 1, \ldots.$$

Introducing the following notation

$$\begin{aligned} &Y^f_{ijr} := \left(h^f\right)^{-r-1} \sigma^f_{ij\underline{r}}, \; w^f_{jr} := \left(h^f\right)^{-r-1} u_{j\underline{r}}, \; \overset{r}{c^f_{is}} := \overset{r}{b^f_{is}} \left(h^f\right)^{\underline{s} - \underline{r}} (s > r), \\ &\overset{r}{Y^f_j} := \left(h^f\right)^{-r-1} \Phi^f_{j\underline{r}}, \quad i,j = \overline{1,3}, \quad s, r = 0, 1, \ldots, \end{aligned} \tag{3.32}$$

equations (3.31) get the following form

$$Y^f_{\alpha jr,\alpha} + \sum_{s=r+1}^{\infty} \overset{r}{c^f_{is}} Y^f_{ijs} + \overset{r}{Y^f_j} = \rho^f \ddot{w}^f_{jr}, \quad j = \overline{1,3}, \quad r = 0, 1, \ldots, \tag{3.33}$$

Taking into account (3.32), from (3.25) we have

$$Y^f_{ijr} = -\delta_{ij}q_r(x_1,x_2,t) + \lambda^f\delta_{ij}\dot{k}^f_r + 2\mu^f\dot{k}^f_{ijr}, \quad i,j=\overline{1,3}, \quad r=0,1,\ldots, \tag{3.34}$$

where $q_r := (h^f)^{-r-1}p_{\underline{r}}$, $\dot{k}^f_{ijr} := (h^f)^{-r-1}\dot{e}^f_{ij\,\underline{r}}$, $\dot{k}^f_r := \dot{k}^f_{iir}$.

Substituting (3.34) into (3.33), by virtue of

$$\dot{e}^f_{\alpha\beta r} = \frac{1}{2}\left(\dot{u}^f_{\alpha r,\beta} + \dot{u}^f_{\beta r,\alpha}\right) + \frac{1}{2}\sum_{s=0}^{r}\left(\overset{r}{a}{}^f_{\beta s}\,\dot{u}^f_{\alpha s} + \overset{r}{a}{}^f_{\alpha s}\,\dot{u}^f_{\beta s}\right)$$

$$-\frac{1}{2}\left[\overset{(\dot{+})}{u}{}^f_{\alpha}\,\overset{(+)}{h}{}^f_{,\beta} - (-1)^r\,\overset{(\dot{-})}{u}{}^f_{\alpha}\,\overset{(-)}{h}{}^f_{,\beta}\right] - \frac{1}{2}\left[\overset{(\dot{+})}{u}{}^f_{\beta}\,\overset{(+)}{h}{}^f_{,\alpha} - (-1)^r\,\overset{(\dot{-})}{u}{}^f_{\beta}\,\overset{(-)}{h}{}^f_{,\alpha}\right],$$

$$\dot{e}^f_{\alpha 3r} = \dot{e}^f_{3\alpha r} = \frac{1}{2}\dot{u}^f_{3r,\alpha} + \frac{1}{2}\sum_{s=0}^{r}\left(\overset{r}{a}{}^f_{\alpha s}\,\dot{u}^f_{3s} + \overset{r}{a}{}^f_{3s}\,\dot{u}^f_{\alpha s}\right)$$

$$+\frac{1}{2}\left[\overset{(\dot{+})}{u}{}^f_{\alpha} - (-1)^r\,\overset{(\dot{-})}{u}{}^f_{\alpha}\right] - \frac{1}{2}\left[\overset{(\dot{+})}{u}{}^f_{3}\,\overset{(+)}{h}{}^f_{,\alpha} - (-1)^r\,\overset{(\dot{-})}{u}{}^f_{3}\,\overset{(-)}{h}{}^f_{,\alpha}\right], \quad \alpha,\beta=1,2, \quad r=0,1,2,\ldots,$$

$$\dot{e}^f_{33r} = \sum_{s=0}^{r} a^f_{3sr}\dot{u}^f_{3s} + \overset{(\dot{+})}{u}{}^f_{3} - (-1)^r\,\overset{(\dot{-})}{u}{}^f_{3}, \quad r=0,1,2,\ldots,$$

where $\overset{(\dot{+})}{u}_j$ and $\overset{(\dot{-})}{u}_j$ are assumed to be known on the face surfaces, we get the system we are looking for. This system after restriction to $r=\overline{0,N}$, for $N=0,1,\ldots$, together with the equations

$$\dot{u}^f_{\alpha r,\alpha} + \sum_{s=0}^{r}\overset{r}{a}{}^f_{is}\dot{u}^f_{is} + \dot{U}^f_r = 0, \quad r=\overline{0,N},$$

where $U^f_r := \overset{(+)}{u}{}^f_3 - \overset{(+)}{u}{}^f_\alpha \overset{(+)}{h}{}^f_{,\alpha} + (-1)^r\left[\overset{(-)}{u}{}^f_\alpha \overset{(-)}{h}{}^f_{,\alpha} - \overset{(-)}{u}{}^f_3\right]$, $r=\overline{0,N}$, gives the desired hierarchical models.

Remark 3.2 In the same manner we can construct hierarchical models when on one face surface either the surface forces or neither the surface forces nor velocities are prescribed, while on the another one the velocities are prescribed.

3.8 Hierarchical Models for Elastic Solid–Fluid Structures

Let us now consider a body consisting of a thin prismatic shell-like elastic solid and fluid parts. In the solid part we use one of the above-mentioned versions of Vekua's hierarchical models for the *N*th approximation, in the fluid part we use one of the above-constructed models in the case of the Nth approximation. As transmission conditions on the cylindrical interface I (with a normal n directed from the solid part to the fluid one) separating solid and fluid parts we take

$$\dot{u}^s_{jr}\big|_I = \dot{u}^f_{jr}\big|_I, \quad \sigma^s_{njr}\big|_I = \sigma^f_{njr}\big|_I, \quad j = \overline{1,3}, \quad r = \overline{0,N}, \tag{3.35}$$

where by the indices f and s, quantities in the fluid and solid parts are indicated; $-\sigma^s_{njr}$ and σ^f_{njr} are the components of the stress vectors acting on the interface from the solid and fluid parts. In the case of the ideal fluid the first of (3.35) should be replaced by the conditions $\dot{u}^s_{nr}\big|_I = \dot{u}^f_{nr}\big|_I, r = \overline{0,N}$, where $\dot{u}^s_{nr}$ and $\dot{u}^f_{nr}$ are the moments of the normal components of the velocity vector on I. Initial conditions have the following form

$$u_{jr}\big|_{t=0} = \varphi_{jr}, \quad \dot{u}_{jr}\big|_{t=0} = \psi_{jr}, \quad j = \overline{1,3}, \quad r = \overline{0,N}, \quad \forall (x_1, x_2) \in \omega := \omega^s \cup \omega^f,$$

where φ_{jr} and ψ_{jr} are prescribed functions. So, we get hierarchical models for structures occupying prismatic domains and consisting of elastic solid and fluid parts.

References

N. Chinchaladze, G. Jaiani, *Hierarchical Mathematical Models for Solid–Fluid Interaction Problems* (Georgian). Materials of the International Conference on Non-Classic Problems of Mechanics, Kutaisi, Georgia, 25–27 October, Kutaisi, vol. 2 (2007), pp. 59–64

G.V. Jaiani, Application of Vekua's dimension reduction method to cusped plates and bars. Bull. TICMI **5**, 27–34 (2001a)

G. Jaiani, On a mathematical model of bars with variable rectangular cross-sections. ZAMM Z. Angew. Math. Mech. **81**(3), 147–173 (2001b)

M. Lucchesi, Longitudinal oscillations of bi-modular rods. Int. J. Struct. Stab. Dyn. **5**(1), 37–54 (2005)

I.N. Vekua, On one method of calculating of prismatic shells (Russian). Trudy Tbilis. Mat. Inst. **21**, 191–259 (1955)

I.N. Vekua, Theory of shallow shells of variable thickness (Russian). Trudy Tbilis. Mat. Inst. **30**, 5–103 (1965)

I.N. Vekua, *Shell Theory: General Methods of Construction* (Pitman Advanced Publishing Program, Boston, 1985)

Chapter 4
Cusped Shells and Plates

Abstract The present chapter is devoted to a survey concerning BVPs for elastic cusped shells, prismatic shells and plates. Researches are carried out within the framework of hierarchical models and classical bending models. Cusped orthotropic plates and cusped plates on an elastic foundation are explored. Vibration problems are studied. Internal concentrated contact interactions in elastic cusped prismatic shell-like bodies are also analyzed. Fundamental conclusions characterizing peculiarities depending on geometry of sharpening of cusped edges are made. $N = 0$ approximation with plane stress, generalized plane stress, and plane deformation, and $N = 1$ approximation with Kirchoff-Love plate model are compared.

Keywords Classical and nonclassical BVPs · Cusped prismatic shells · Cusped plates · Cusped cylindrical and conical shells · Vibration

4.1 First Investigations: Fundamental Statement

Bisshopp (1944) constructed solutions for lateral bending of a symmetrically loaded elastic conical disc (see Fig. 4.1) with the thickness

$$h_m = h_0\left(1 - \frac{r}{r_0}\right), \quad 0 \leq r := \left(x_1^2 + x_2^2\right)^{1/2} < r_0, \quad h_0,\, r_0 = const > 0.$$

In the polar coordinates the bending equation has the following form

G. Jaiani, *Cusped Shell-Like Structures*, SpringerBriefs in Applied Sciences and Technology, DOI: 10.1007/978-3-642-22101-9_4,

Fig. 4.1 A cross-section of the conical disc

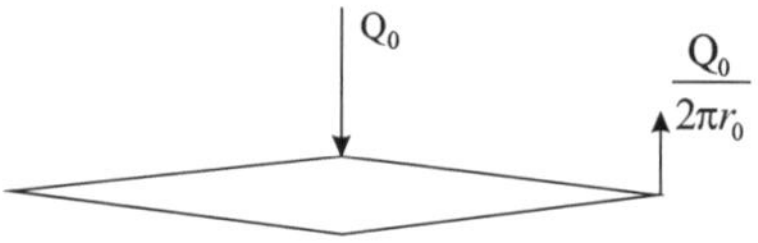

$$D_M\nabla^2\nabla^2 w + \frac{dD_M}{dr}\Big(2\frac{\partial^3 w}{\partial r^3} + \frac{2+\nu}{r}\frac{\partial^2 w}{\partial r^2} - \frac{1}{r^2}\frac{\partial w}{\partial r} + \frac{2}{r^2}\frac{\partial^3 w}{\partial r\partial\theta^2} - \frac{3}{r^3}\frac{\partial^2 w}{\partial\theta^2}\Big) + \frac{d^2 D_M}{dr^2}\Big(\frac{\partial^2 w}{\partial r^2} + \frac{\nu\partial w}{r\partial r} + \frac{\nu}{r^2}\frac{\partial^2 w}{\partial\theta^2}\Big) = q, \tag{4.1}$$

where $D_M := 12^{-1}Eh_M^3(1-\nu^2)^{-1}$ is the flexural rigidity of the plate, ν is Poisson's ratio, E is Young's modules of elasticity, q is a lateral load,

$$\nabla^2 := \frac{\partial^2}{\partial r^2} + \frac{1}{r}\frac{\partial}{\partial r} + \frac{1}{r^2}\frac{\partial^2}{\partial\theta^2}.$$

As we see, the edge $r = r_0$ is a cusped one. Expressions for classical bending moments M_r, M_θ, and deflections w are found by Bisshopp in hyper-geometrical functions and quadratures, respectively. In 1955 for the more general case with the flexural rigidity

$$D = D_0\left[1-\left(\frac{r}{r_0}\right)^{\alpha_0}\right]^{\beta_0}, \quad 0\le r<r_0, \quad D_0, r_0, \alpha_0, \beta_0 = \text{const} > 0, \tag{4.2}$$

Kovalenko (1959) constructed solutions of symmetrical bending of elastic discs (pp. 41–105, and references therein). But he solved only one problem for the cusped disc considered by Bisshopp ($\alpha_0 = 1, \beta_0 = 3$), namely, when cusped edge is "simply supported" (according to the author) and at the center of the disc a concentrated force Q_0 is applied [see Kovalenko (1959), p. 104]. However, this problem is mathematically non-well-posed, since (see Fig. 4.1)

$$M_r|_{r=r_0} = 0, w_r|_{r=r_0} = \infty, Q_r|_{r=r_0} = -\frac{Q_0}{2\pi r_0}.$$

As we see below, the above edge may be simply supported only if $\beta_0<3$ and, moreover, it will be correctly set only if $\beta_0<2$, since if $2\le\beta_0<3$ and $M_r|_{r=r_0}\ne 0$, the problem will not be solvable.

The first correct work concerning classical bending of cusped elastic plates was done by Makhover and Mikhlin [see Mikhlin (1970)]. The bending equation of the Kirchhoff-Love plate of variable thickness has the form [see e.g. Jaiani (2002)]

$$Bw := (Dw_{,11})_{,11} + (Dw_{,22})_{,22} + \nu(Dw_{,22})_{,11} + \nu(Dw_{,11})_{,22} + 2(1-\nu)(Dw_{,12})_{,12} = q(x_1,x_2), \tag{4.3}$$

where w is a deflection, $D := \frac{2Eh^3}{3(1-\nu^2)}$ is a flexural rigidity of the plate, q is the lateral load. Using the results of Mikhlin, Makhover (1957, 1958) considered such a cusped plate with an adjacent to the axis x_1 bounded projection ω lying in the half-plane $x_2 \geq 0$ and the flexural rigidity $D(x_1, x_2)$ satisfying

$$D' x_2^{k_1} \leq D(x_1, x_2) \leq D'' x_2^{k_1}, \quad x_2 \geq 0, \quad D', D'', k_1 = \text{const} > 0. \tag{4.4}$$

Namely, she has shown that for $k_1 < 2$ the deflection can be prescribed on the cusped edge $x_2 = 0$ of the plate, while its normal derivative can be prescribed for $k_1 < 1$ there. Applying more natural spaces than used by Makhover, Jaiani (1987a, 1984a, 1976) has analyzed in which cases the cusped edge can be fixed ($k_1 < 1$) or simply supported ($k_1 < 2$). Moreover, he gave the correct formulation of all the reasonable principal BVPs. The corresponding BVPs are solved in the explicit (integral) form by Jaiani (1977), when ω is the half-plane.

In the static case from (3.1), (3.11), (3.12), (3.6–3.9), if $N = 0$ we immediately get the governing system of the $N = 0$ approximation (superscript $N = 0$ is omitted below)

$$-\mu\left[(h v_{\alpha 0,\beta})_{,\alpha} + (h v_{\beta 0,\alpha})_{,\alpha}\right] - \lambda (h v_{\gamma 0,\gamma})_{,\beta} = \overset{0}{X}_\beta, \quad \beta = 1, 2, \tag{4.5}$$

$$-\mu (h v_{30,\alpha})_{,\alpha} = \overset{0}{X}_3, \tag{4.6}$$

and the relations

$$X_{\alpha j0,\alpha} + \overset{0}{X}_j = 0, \tag{4.7}$$

$$\begin{aligned} e_{\alpha\beta 0} &= \frac{1}{2}(u_{\alpha 0,\beta} + u_{\beta 0,\alpha}) - \frac{1}{2}(\ln h)_{,\alpha}\, u_{\beta 0} - \frac{1}{2}(\ln h)_{,\beta}\, u_{\alpha 0} \\ &= \frac{h}{2}(v_{\alpha 0,\beta} + v_{\beta 0,\alpha}), \end{aligned} \tag{4.8}$$

$$e_{\alpha 3} = \frac{1}{2} u_{30,\alpha} - \frac{1}{2}(\ln h)_{,\alpha}\, u_{30} = \frac{h}{2} v_{30,\alpha}, \quad e_{330} = 0, \tag{4.9}$$

$$X_{\alpha j0} = \lambda \delta_{\alpha j} e_{\gamma\gamma 0} + 2\mu e_{\alpha j0}, \tag{4.10}$$

$$X_{330} = \nu X_{\alpha\alpha 0}, \tag{4.11}$$

$$(u_i, X_{ij}, e_{ij})(x_1, x_2, x_3) = \frac{1}{2h}(u_{i0}, X_{ij0}, e_{ij0})(x_1, x_2), \tag{4.12}$$

$$(u_{i0}, X_{ij0}, e_{ij0})(x_1, x_2, x_3) = \int\limits_{\overset{(-)}{h}(x_1, x_2)}^{\overset{(+)}{h}(x_1, x_2)} (u_i, X_{ij}, e_{ij})(x_1, x_2, x_3)\, dx_3, \tag{4.13}$$

$$u_i(x_1,x_2,x_3) = \frac{1}{2}v_{i0}(x_1,x_2) = \frac{1}{2h}u_{i0}(x_1,x_2). \tag{4.14}$$

Let us now compare Vekua's zero approximation with the plane stress, generalized plane stress, and the plane strain. To this end let $2h(x_1,x_2) = \text{const}$. As we see from (4.12), (4.13), like the generalized plane stress, in Vekua's zero approximation instead of u_i, X_{ij}, e_{ij} their integral (with respect to x_3) mean values are considered but in contrast to the plane stress and generalized plane stress

$$X_{i3}\big|_{x_3=\pm h} \not\equiv 0, \quad i = \overline{1,3}.$$

Moreover, like the plane strain we have (4.11); that is why, if $2h = \text{const}$, (4.5) coincides with the plane strain system:

$$\mu u_{\beta 0,\alpha\alpha} + (\lambda+\mu)u_{\gamma 0,\gamma} + \overset{0}{X}_\beta = 0, \quad \beta = 1,2.$$

In contrast to the classical plane strain and generalized plane stress $u_3 \not\equiv 0$[1] and, by virtue of (4.14), we have equation (4.6) for u_3. It is easy to verify that under assumptions of the plane strain ($u_3 \equiv 0$, u_α, $\alpha = 1,2$, are independent of x_3), by virtue of (4.12), (4.13), from (4.7)–(4.14) there follow the relations of the plane strain. In this sense Vekua's zero model is consistent with the plane strain. Neither assumptions of the plane stress ($X_{i3} = 0$, $i = \overline{1,3}$) nor assumptions of the generalized plane stress ($X_{i3}\big|_{x_3=\pm h} = 0$, $i = \overline{1,3}$, $X_{n3} = 0$ on the lateral surface, and $\Phi_3 = 0$) lead to such consistency with Vekua's zero model, since in the cases of both the plane stress and generalized plane stress in the corresponding to (4.10) relation instead of λ arises

$$\lambda^* = \frac{2\lambda\mu}{\lambda + 2\mu}.$$

In the zero approximation of I.Vekua's hierarchical models of shallow prismatic shells Khvoles (1971) represented the forth order degenerate operator acting on the Airy stress type function as a product of two second order degenerate operators in the case when the plate thickness $2h$ is given by

$$2h = h_0 x_2^{k_2}, \quad h_0 = \text{const} > 0, \quad k_2 = \text{const} > 0, \quad x_2 \geq 0, \tag{4.15}$$

and investigated the question of the general representation of corresponding solutions. In the case (4.15) Jaiani (1972, 1980a, 1988) investigated the tension-compression problem of cusped plates, based on the zero approximation of I. Vekua's hierarchical models of shallow prismatic shells [see system (4.5), (4.6)] and constructed [see (1973, 1974b, 1975, 1980b, 1982)] effective solutions of

[1] In the case of the generalized plane stress u_3 is so small that it can be assumed to be equal to zero but the same cannot be supposed concerning its derivative $u_{3,3} = e_{33}$. Meanwhile, in view of (4.9), (4.12), the strain tensor component $e_{33} \equiv 0$ in Vekua's zero approximation.

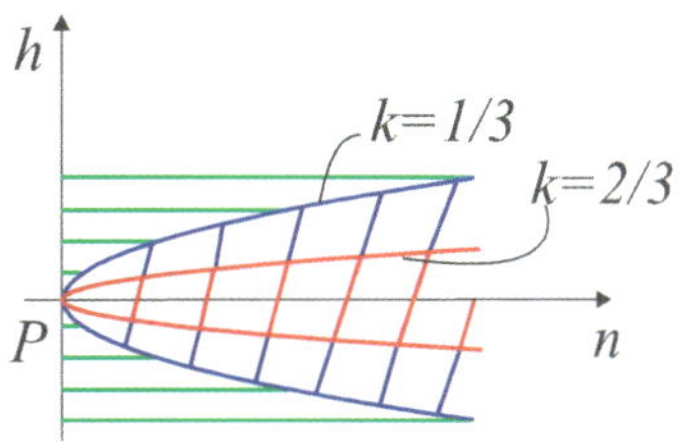

Fig. 4.2 Zones of setting of different BCs at cusped edges depending on geometry of tapering (*sharpening*)

BVPs in integrated stresses, when ω is the half-plane, half-strip, or infinite plane sector, using, found to this end, singular solutions depending only on the polar angle with a shift. The well-posedness of BVPs for systems (3.14) and (4.5), (4.6) essentially differs. This matter will be discussed in the forthcoming paper of G. Jaiani.

A qualitative part of G. Jaiani's above-mentioned results can be summarized as follows.

Statement 4.1 *Let n be the inward normal of the prismatic shell projection boundary. In the case of the tension-compression* ($N = 0$) *problem on the cusped edge, where* (*see* Figs. 2.25 *and* 2.31, 2.32, 2.33, 2.34, *which can be considered as cusped prismatic shells' cross-sections*) $0 \leq \frac{\partial h}{\partial n} < \infty$ (*in the case* (4.15) *this means* $k_2 \geq 1$), *which will be called a sharp cusped edge, the displacement vector can not be prescribed, while on the cusped edge, where* (*see* Figs. 2.26, 2.27, 2.28, 2.29, 2.30) $\frac{\partial h}{\partial n} = +\infty$ [*in the case* (4.15) *this means* $k_2 < 1$], *which will be called a blunt cusped edge, the displacement vector can be prescribed. In the case of the classical bending problem when at a plate cusped edge*

$$\frac{\partial h}{\partial n} = O(d^{k-1}) \quad \text{as} \quad d \to 0, \quad k = \text{const} > 0, \tag{4.16}$$

where d is the distance between an interior reference point of the plate projection and the cusped edge, the edge can not be fixed if $k \geq 1/3$, *but it can be fixed if* $0 < k < 1/3$; *it can not be simply supported if* $k \geq 2/3$, *and it can be simply supported if* $0 < k < 2/3$; *it can be free or arbitrarily loaded by a shear force and a bending moment if* $k > 0$. *Note that in the case* (4.15), *the condition* (4.16) *implies that* $d = x_2$ and $k = k_2 = k_1/3$. *In other words, when a profile, i.e., cross-section* (*see* Fig. 4.2) *of the plate in a neighborhood of a plate boundary point P* (*blunt cusp*):

1. *lies in the green zone then all three main BVPs can be correctly posed;*
2. *lies in the blue zone then the edge can be either simply supported or be free* (*or be arbitrarily loaded*);
3. *lies in the red zone then such edge can be only free* (*or be arbitrarily loaded*). *The last from the above three assertions is also valid for the sharp cusp.*

4.2 Cusped Cylindrical and Conical Shells

It was to be expected that the above conclusions of Statement 4.1 remain true also in the case of cusped standard shells. In concrete cases of cusped cylindrical and conical shells bending, it has been shown by Tsiskarishvili and Khomasuridze (1991a, b) [see also Tsiskarishvili (1993)]. The cylindrical bending of a cylindrical shell has been considered when the thickness has the following form

$$h = h_0 \sin^{\kappa} \varphi, \quad h_0, \kappa = \text{const} > 0,$$

where $\varphi \in]0, \varphi_0[$ is the polar angle counted from the plane $O\zeta x$ (see Fig. 4.3). Let w be a radial component of the displacement vector, i.e., deflection, which obviously characterizes bending, and v be an angular component of the displacement vector, i.e., displacement in the middle surface (cylinder) orthogonal to the generatrix of the cylinder. The displacement v characterizes tension-compression of the cylindrical shell. They proved (as foreseen by G. Jaiani) that deflection w can be given on the cusped edge only if

$$\kappa \in \left]0, \frac{2}{3}\right[,$$

its first derivative can be given only if

$$\kappa \in \left]0, \frac{1}{3}\right[,$$

but v can be given if the cusped edge is blunt, i.e., $\kappa \in]0, 1[$, as in the zero order approximation of I.Vekua's version which characterizes tension-compression (compare with Statement 4.1). In the case under consideration, the remarkable effect is the following: on the one hand, as it is well-known, the cylindrical bending of a cylindrical shell is always accompanied by significant tension-compression. Therefore, w can not be separated from v by neglecting the latter. On the other hand, they have conserved their properties characterizing them correspondingly by bending and tension-compression of cusped plates when these kinds of deformations can be considered separately. The same authors have considered the strength problem of a uniformly loaded $3D$ elastic structure consisting of a rotational cylindrical shell of constant thickness and two rotational cusped conical shells with a linearly changing thickness

$$h_{con} := h_0 x_1 tg\alpha, \quad h_0, \; \alpha = \text{const} > 0,$$

[see Fig. 4.4, where an axial (z is the axis of symmetry of the rotational body) section of the 3D structure is given]. The conical shell is considered in coordinates x_1, n_1, where x_1 is taken along an generatrix of the conical shell middle surface and n_1 is orthogonal to x_1; the origin of the coordinate system is taken at the cone vertex. The cylindrical shell is considered in coordinates x_2, n_2, where x_2 is taken along the generatrix of the cylindrical shell middle surface and n_2 is orthogonal to x_2;

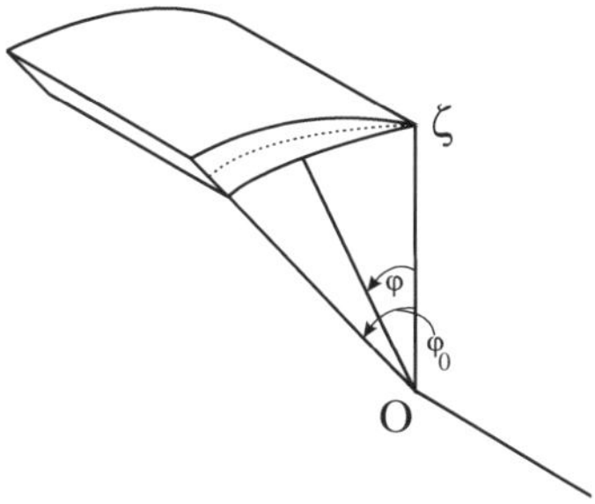

Fig. 4.3 A cusped cylindrical shell

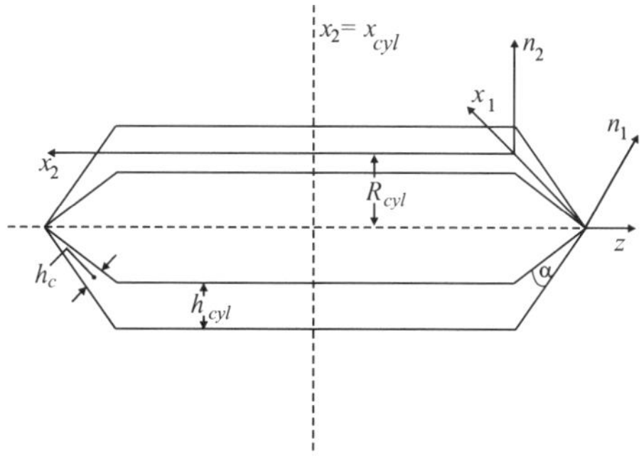

Fig. 4.4 Cross-section of the 3D cusped structure

the origin of the coordinate system is taken at an edge of the cylindrical shell. The aim is to choose appropriately the angle of sharpening of the conical shell in order to guarantee the required strength of the structure. The plane $x_2 = x_{cyl} = \text{const}$ is the plane of symmetry of the structure and stress state. Therefore, it is sufficient to consider a half of the structure. Moreover, since z is the other axis of symmetry, the consideration can be restricted to the one fourth of the structure cross-section (see Fig. 4.4). The stress state of the structure is determined under symmetry

$$w_{cyl}(x_{cyl}) = 0, \quad \frac{\partial u_{cyl}}{\partial x_2}(x_{cyl}) = 0, \quad \frac{\partial^2 u_{cyl}}{\partial x_2^2}(x_{cyl}) = 0$$

and transmission (at the contact of the cylindrical and conical shells' middle surfaces)

$$u_{con} = u_{cyl}, \; w_{con} = w_{cyl}, \; \theta_{con} = \theta_{cyl},$$

$$N_{con} = N_{cyl}, \; M_{con} = M_{cyl}, \; T_{con} = T_{cyl}$$

conditions. Here u_{con}, u_{cyl} are displacements along the middle surface generatrices; w_{con}, w_{cyl} are deflections; $\theta_{con}, \theta_{cyl}$ are tangents of rotation angles; $N_{con}, N_{cyl}, T_{con}, T_{cyl}$ are normal and tangent forces; M_{con}, M_{cyl} are bending moments and the subscripts "con" and "cyl" mean conical and cylindrical shells, respectively.

4.3 Cusped Plates Bending

It was also to be expected that the assertions of Statement 4.1 do not depend on anisotropy of material of elastic body. Example, consider the case of a classical bending of an orthotropic cusped plates with a governing equation

$$\begin{aligned} Jw :=&(D_1 w_{,11})_{,11} + (D_2 w_{,22})_{,22} + (D_3 w_{,22})_{,11} + (D_3 w_{,11})_{,22} \\ &+ 4(D_4 w_{,12})_{,12} = q(x_1, x_2) \ \ in \ \ \omega \subset \mathbb{R}^2, \end{aligned} \tag{4.17}$$

where w is a deflection; q is a lateral load; $D_i \in C^2(\omega)$, $i = \overline{1,4}$, and

$$D_i := \frac{2E_i h^3}{3}, \quad i = \overline{1,3}, \quad D_4 := \frac{2Gh^3}{3};$$

$$D_\alpha - D_3 > 0, \quad \alpha = 1, 2 \quad \text{if} \ \ h > 0;$$

E_i, $i = \overline{1,3}$, and G are elastic constants for the orthotropic case.

In particular, if the plate is isotropic,

$$E_\alpha = \frac{E}{1 - \nu^2}, \quad \alpha = 1, 2, \ \ E_3 = \frac{\nu E}{1 - \nu^2}, \quad G = \frac{E}{2(1 + \nu)},$$

and [see (4.3)]

$$Jw \equiv Bw.$$

Let $\partial\omega$ be the piecewise smooth boundary of the domain ω with a part γ_0 lying on the axis Ox_1 and a part γ_1 lying in the half-plane $x_2 > 0$ ($\partial\omega \equiv \overline{\gamma}_0 \cup \overline{\gamma}_1$).

Let further the thickness $2h > 0$ in $\omega \cup \gamma_1$, and $2h \geq 0$ on γ_0. Therefore,

$$D_i(x_1, x_2) > 0 \ \ in \ \ \ \omega \cup \gamma_1, \quad D_i(x_1, x_2) \geq 0 \text{ on } \gamma_0, \quad i = \overline{1,4}.$$

In the particular case when [compare with (4.4)]

$$D_{1i} x_2^\varkappa \leq D_i(x_1, x_2) \leq D_{2i} x_2^\varkappa, \ \ i = \overline{1,4}, \ \ \text{in} \ \ \omega,$$

where

$$D_{\alpha i} = \text{const} > 0, \quad \alpha = 1, 2, \quad i = \overline{1,4}, \ \varkappa = \text{const} \geq 0,$$

i.e.,

$$D_i(x_1, x_2) = \widetilde{D}_i(x_1, x_2) x_2^\varkappa, \quad D_{1i} \leq \widetilde{D}_i(x_1, x_2) \leq D_{2i},$$

$$D_{1\alpha} > D_{23}, \quad \alpha = 1, 2,$$

Statement 4.1 remains valid [see (Jaiani1999a)]. The vibration problem of such a cusped orthotropic plate is analyzed in Jaiani (1999b).

We recall that

$$\begin{aligned}
M_\alpha &= -(D_\alpha w_{,\alpha\underline{\alpha}} + D_3 w_{,\beta\underline{\beta}}), \quad \alpha \neq \beta, \quad \alpha, \beta = 1, 2,\\
M_{12} &= -M_{21} = 2D_4 w_{,12},\\
Q_\alpha &= M_{\alpha,\underline{\alpha}} + M_{21,\beta}, \quad \alpha \neq \beta, \quad \alpha, \beta = 1, 2,\\
Q_\alpha^* &= Q_\alpha + M_{21,\beta}, \quad \alpha \neq \beta, \quad \alpha, \beta = 1, 2,
\end{aligned}$$

where M_α are bending moments, $M_{\alpha\beta}$, $\alpha \neq \beta$, are twisting moments, Q_α are shear forces and Q_α^* are generalized shear forces. At points of the plate boundary, where the thickness vanishes, all quantities will be defined as limits from inside ω.

Let

$$I_{ki} = I_{ki}(x_1) = \int_0^{l(x_1)} x_2^k D_i^{-1}(x_1, x_2) dx_2, \ (x_1, 0) \in \gamma_0, \ (x_1, l(x_1)) \in \omega, \quad i = \overline{1,4},$$
$$k = \overline{0,2}.$$

The BVPs for equation (4.17) are correct under the following BCs: on γ_1

$$w = g_{12}, \quad \frac{\partial w}{\partial n} = g_{22},$$

and on γ_0 either

$$w = g_{11}, \ w_{,2} = g_{21} \quad \text{if} \quad I_{0i} < +\infty \ (0 < \varkappa < 1),$$

or

$$w_{,2} = g_{21}, \ Q_2^* = h_2 \quad \text{if} \quad I_{0i} < +\infty \ (0 < \varkappa < 1),$$

or

$$w = g_{11},$$

$$(M_2 w) = h_1 \begin{cases} \not\equiv 0 & \text{when} \quad I_{0i} < +\infty \ (0 \leq \varkappa < 1), \\ \equiv 0 & \text{when} \quad I_{0i} = +\infty \ (1 \leq \varkappa < 2) \end{cases}$$

if $I_{2i} < +\infty \quad (0 \leq \varkappa < 3)$, or

$$(M_2 w) = h_1 \begin{cases} \not\equiv 0 \text{ when} & I_{0i} < +\infty \quad (0 \leq \varkappa < 1), \\ \equiv 0 \text{ when} & I_{0i} = +\infty \quad (1 \leq \varkappa < +\infty), \end{cases}$$
$$(Q_2^* w) = h_2 \begin{cases} \not\equiv 0 \text{ when} & I_{1i} < +\infty \ (0 \leq \varkappa < 2), \\ \equiv 0 \text{ when} & I_{1i} = +\infty \ (2 \leq \varkappa < +\infty) \end{cases}$$

if $I_{0i} \leq +\infty$ $(0 \leq \varkappa < +\infty)$, where $g_{\alpha\beta}$, h_α, $\alpha, \beta = 1, 2$ are prescribed functions.

However, for general cusped shells and also for general anisotropic cusped plates, corresponding to Statement 4.1 analysis is yet to be done.

The classical bending of plates with the flexural rigidity (4.4) in energetic and in weighted Sobolev spaces has been studied by Jaiani (1992, 1996, 2002) [see also Jaiani and Kufner (2006)]. In the energetic space some restrictions on the lateral load has been relaxed by Devdariani (1992). Using Kovalenko's (1959) above-mentioned representations for deflections, radial derivatives of deflections, bending moments, and shear forces, Tsiskarishvili (1993) characterized completely the classical axial symmetric bending of circular cusped plates without or with a hole, when the governing equation and the flexural rigidity have the forms (4.1) and (4.2), respectively.

Chinchaladze (2002d) for the Kirchhoff-Love plate with two cusped edges $x_2 = 0,\ x_2 = l$ (the flexural rigidity has the form

$$D(x_2) = D_0 x_2^{\alpha}(l - x_2)^{\beta}, \quad l, D_0, \alpha, \beta = \text{const} > 0)$$

has explored the well-posedness of Keldysh' type and weighted BVPs in the case of the cylindrical bending; reasonable dynamical problems are also investigated; general and harmonic vibration of such plates are studied; it is shown that the setting of BCs at the plate's edges depends on the geometry of sharpening of plate edges, while the setting of initial conditions is independent of it; in some cases the solutions of these problems are represented explicitly by either integrals or series.

Bending of an elastic cusped Kirchhoff-Love plate on an elastic foundation with a compliance

$$k'' \geq k(x_1, x_2) \geq k' > 0, \quad k', k'' = \text{const},$$

was studied by Jaiani (2004b) in weighted Sobolev spaces. The governing equation has the following form

$$Bw + k(x_1, x_2)w = q.$$

The operator B is defined in (4.3). The posing of BCs depends on convergence-divergence of the integrals

$$\int_0^{\varepsilon} \tau^{\alpha} D^{-1}(x_1, \tau)d\tau, \quad \alpha = 1, 2,$$

in a one-sided $\varepsilon > 0$ neighborhood of the plate cusped edge γ_0. For geometrically non-linear cusped Timoshenko plate with flexural rigidity (4.3) we have only one result of Jaiani and Chinchaladze (2009): at a cusped edge the bending moment and shear force can be prescribed only if $0 \leq k_1 < 2$ and $0 \leq k_1 < 1$, respectively [see (4.4)] in contrast to the linear case, when these quantities can always be prescribed.

4.4 The *N*th Order Approximation

As it has been already mentioned the problems involving cusped prismatic shells, in particular plates, lead to the problem of correct mathematical formulations of BVPs for even order elliptic equations and systems whose orders degenerate on the boundary. The works of Jaiani (1988, 1974a, 1990, 1987b, 1999c, 1984b) are devoted to investigation of BVPs for more general classes of equations and systems than the above-mentioned ones.

Applying the function-analytic method developed by Fichera (1956, 1960), the particular case $\lambda = \mu$ of Vekua's system (4.5), (4.6) of the $N = 0$ approximation for general form cusped prismatic shells has been investigated by Jaiani (1988). The main conclusion says that at blunt cusped edge $\left(\frac{\partial h}{\partial n} = +\infty\right)$ displacement vector components can be prescribed, while sharp cusped edge $\left(0 \leq \frac{\partial h}{\partial n} < \infty\right)$ should be freed from BCs (Keldysh type BVP for displacements). The last result concerning sharp cusped edges is true for the Nth approximation as well [see Jaiani (2001)].

As it follows from (3.11), when $N = 1$, in the static case for the symmetric prismatic shell I. Vekua's system in the $N = 1$ approximation has the form

$$- \mu\left[(hv_{\alpha 0,\beta})_{,\alpha} + (hv_{\beta 0,\alpha})_{,\alpha}\right] - \lambda(hv_{\gamma 0,\gamma})_{,\beta} \tag{4.18}$$

$$- 3\lambda(hv_{31})_{,\beta} = \overset{0}{X}_\beta, \quad \beta = 1,2,$$

$$- \mu(hv_{30,\alpha})_{,\alpha} - 3\mu(hv_{\alpha 1})_{,\alpha} = \overset{0}{X}_3, \tag{4.19}$$

$$- 3\mu\left[(h^3 v_{\alpha 1,\beta})_{,\alpha} + (h^3 v_{\beta 1,\alpha})_{,\alpha}\right] - 3\lambda\left(h^3 v_{\gamma 1,\gamma}\right)_{,\beta} \tag{4.20}$$

$$+ 3\left[\mu h(v_{30,\beta} + 3v_{\beta 1})\right] = 3h\overset{1}{X}_\beta, \quad \beta = 1,2,$$

$$- 3\mu(h^3 v_{31,\alpha})_{,\alpha} + 3\left[\lambda hv_{\gamma 0,\gamma} + 3(\lambda + 2\mu)hv_{31}\right] = 3h\overset{1}{X}_3. \tag{4.21}$$

In the case (4.15) the tension-compression system (4.18), (4.21) is investigated by Devdariani et al. (2000) [see also Devdariani (2001)]. The existence and uniqueness of generalized solutions of BVPs with Dirichlet (for weighted zero-moments when $k_2 < 1$ and for weighted first moments when $k_2 < 1/3$) and Keldysh type (for weighted zero-moments when $k_2 \geq 1$ and for weighted first moments when $k_2 \geq 1/3$) BCs is proved in weighted Sobolev spaces.

In Jaiani and Schulze (2007) the vibration tension-compression system (vanishing of the vibration frequency corresponds to the static system) is investigated under all reasonable nonhomogeneous Dirichlet, weighted Neumann, and mixed BCs when the thickness satisfies the unilateral condition

$$2h(x_1, x_2) \geq h_\kappa x_2^\kappa, \quad h_\kappa = \text{const} > 0, \quad \kappa = \text{const} \geq 0, \quad x_2 \geq 0.$$

The bending [see system (4.19), (4.20)] vibration problem can be investigated in an analogous manner. Factors 3 in (4.20), (4.21) guarantee coerciveness of the bilinear forms corresponding to the bending and tension-compression systems.

Note that for the plate of a constant thickness $2h$ from the bending system (4.19), (4.20), under the corresponding assumptions we can derive the classical bending equation

$$\Delta\Delta u_3 = \frac{q}{D^*},$$

where

$$u_3 = \frac{1}{2} v_{30}$$

and

$$D^* = \frac{(1-\nu)^2}{1-2\nu} D.$$

Here D is the classical flexural rigidity. So, Vekua's plate bending model in the $N = 1$ approximation actually coincides with the classical bending model but by bending Vekua's plate is flexurally more rigid than classical one [see e.g. Jaiani (2004a), pp. 210–212].

The method of investigation of hierarchical models based on the idea to get Korn's type inequality for 2D models from the 3D Korn's inequality for non-cusped domains and then to use Lax-Milgram theorem belongs to Gordeziani (1974a, 1974b) which (the method) found its complete realization in Avalishvili and Gordeziani (2003). The analogous approach is developed by Schwab (1996) in connection with the hierarchical models for non-cusped bodies [see also Dauge et al. (1996)] and references therein]. This idea for cusped but Lipschitz 3D domains with corresponding modifications was successfully used by Jaiani et al. (2003, 2004). In the case when cusped prismatic shell occupies a Lipschitz 3D domain, on face surfaces stress vectors, while on the non-cusped edge weighted moments of displacement vector components are given, with the help of variational methods, the existence and uniqueness theorems for the corresponding 2D BVPs are proved in appropriate weighted function spaces endowed with the norm

$$\begin{aligned} ||w||^2_{H^1(\Omega)} = \sum_{i=1}^{3} \Big[& \sum_{\underline{r_i}=0}^{N_i} \big(r_i + \frac{1}{2}\big) ||h^{r_i+\frac{1}{2}} v_{\underline{i}r_i}||^2_{L_2(\omega)} \\ & + \sum_{\underline{s_i}=0}^{N_i} \big(s_i + \frac{1}{2}\big) || \sum_{\underline{r_i}=s_i}^{N_i} \big(r_i + \frac{1}{2}\big) [1 - (-1)^{r_i+s_i}] h^{r_i-\frac{1}{2}} v_{\underline{i}r_i} ||^2_{L_2(\omega)} \\ & + \sum_{\alpha=1}^{2} \sum_{\underline{s_i}=0}^{N_i} \big(s_i + \frac{1}{2}\big) || h^{s_i+1/2} v_{\underline{i}s_i,\alpha} + \sum_{\underline{r_i}=s_i+1}^{N_i} \big(r_i + \frac{1}{2}\big) 2A_{\alpha r_i - s_i} h^{r_i+\frac{1}{2}} v_{\underline{i}r_i} ||^2_{L_2(\omega)} \Big]. \end{aligned}$$

where

$$A_{\alpha q} = -\frac{\overset{(+)}{h}_{,\alpha} - (-1)^q \overset{(-)}{h}_{,\alpha}}{2h}.$$

By means of the solutions of these 2D BVPs, a sequence of approximate solutions in the corresponding 3D region is constructed. This sequence converges in the Sobolev space H^1 to the solution of the corresponding original 3D BVP. The systems of differential equations corresponding to the 2D variational hierarchical models are explicitly constructed for a general system and for Legendre polynomials, i.e., I.Vekua's case, in particular. The above method does not allow to consider BVPs when on the cusped edge either displacements or loads (the loads in this case are concentrated along the cusped edge ones) are prescribed. Similar investigations for plates, prismatic and general standard shells whose thickness may vanish on their boundaries but occupy Lipschitz 3D domains are carried out by Gordeziani et al. (2005, 2006), Avalishvili et al. (2008, 2010), Avalishvili (2004) [see also Miara et al. (2010)]. Gordeziani and Avalishvili (2004, 2005) are devoted to the design of a hierarchy of 2D models for dynamical problems within the theory of multicomponent linearly elastic mixtures in the case of prismatic shells with thickness which may vanish on some parts of its boundary, provided that the 3D domain occupied by the prismatic shell is Lipschitz one.

By Chinchaladze et al. (2008) the well-posedness of BVPs for elastic cusped plates (i.e., symmetric prismatic shells) in the Nth approximation $N \geq 0$ of I.Vekua's hierarchical models [see system (3.11) in the static case] under all the reasonable BCs at the cusped edge and given displacements at the non-cusped edge is studied. The approach works also for non-symmetric prismatic shells word for word. Special attention is drawn to the $N = 0, 1, 2$ approximations as to important cases from the practical point of view. For example, $N = 0$ and $N = 1$ models, roughly speaking, coincide with the plane deformation and Kirchhoff-Love model, respectively. There is assumed that the cusped plate projection ω has a Lipschitz boundary $\partial\omega = \overline{\gamma}_0 \cup \overline{\gamma}_1$, where $\overline{\gamma}_0$ is a segment of the x_1-axis and γ_1 lies in the upper half-plane $x_2 > 0$; moreover, in some neighborhood of an edge of the plate, which may be cusped, the plate thickness has the form (4.15), where $k_2 = \kappa$. Then γ_0 will be a cusped edge for $\kappa > 0$. Note that in the last case, on the one hand, a 3D domain Ω occupied by the plate is non-Lipschitz for $\kappa > 1$; on the other hand, the governing system consisting of $3N + 3$ simultaneous equations [see (3.11) in the static case] is elliptic in Ω and has an order degeneration on γ_0 for any $\kappa > 0$. The classical and weak setting of the BVPs in the case of the Nth approximation is considered. For arbitrary $\kappa \geq 0$ appropriate weighted function spaces $X_N^\kappa(\omega)$ which are crucial in analysis of the problem are introduced. $X_N^\kappa(\omega)$ is the completion of the space $[D(\omega)]^{3N+3}$ with the help of the norm[2]:

$$\|v\|^2_{X_N^\kappa} = (v, v)_{X_N^\kappa} = \sum_{r=0}^{N}\left(r+\frac{1}{2}\right)\int_{\omega}\sum_{i,j=1}^{3} e^2_{ijr}(v)\frac{d\omega}{h} = \frac{1}{4}\sum_{r=0}^{N}\sum_{i,j=1}^{3}\left(r+\frac{1}{2}\right)$$

$$\times \int_{\omega}\left[h^{r+1}\left(v_{ir,j}+v_{jr,i}\right)+\sum_{s=r+1}^{N} h^{s+1}\left(\overset{r}{b}_{js}v_{is}+\overset{r}{b}_{is}v_{js}\right)\right]^2\frac{d\omega}{h},$$

where $\overset{r}{b}_{is}$ are already introduced functions (see Sect. 3.1) depending only on h. Coerciveness of the corresponding bilinear form is shown and uniqueness and

---

[2] $D(\omega)$ is the space of infinitely differentiable function Φ with compact support in ω, i.e., $\Phi = 0$ outside a closed and bounded subset of ω. Generally, the support of a function $\Phi(x_1, x_2)$ is the closure of the set of (x_1, x_2) for which $\Phi(x_1, x_2) \neq 0$.

existence results for the variational problem are proved. The structure of the spaces X_N^κ is described in detail and their connection with weighted Sobolev spaces is established. Moreover, some sufficient conditions for a linear functional arising in the right hand side of the variational equation to be bounded are given. $N = 0, 1, 2$ approximations are considered in detail. Peculiarities characterizing these concrete models are exposed. Note that for the rth order moments Dirichlet and Keldysh BCs are correct when

$$\kappa < \frac{1}{2r+1} \quad \text{and} \quad \kappa \geq \frac{1}{2r+1}, \qquad r = \overline{0, N},$$

respectively.

Jaiani (2008a) deals with a system consisting of singular partial differential equations of the first and second order arising in the zero approximation of I.Vekua's hierarchical models of prismatic shells, when the thickness of the prismatic shell varies as a power function of one argument and vanishes at the cusped edge of the shell [see (4.15)]. For this system of special type a nonlocal BVP in the half-plane is solved in the explicit form. The BVP under consideration corresponds to the stress-strain state of the cusped prismatic shell under the action of concentrated forces and concentrated couples (see also below a footnote on page 46).

4.5 Internal Concentrated Contact Interactions in Cusped Prismatic Shell-Like Bodies

Continuum mechanics envisages two types of interactions between a body and its environment and between body parts, namely, contact interactions and interactions at a distance [see Introduction of Chinchaladze et al. (2011)]. The former are always modeled, after Cauchy, as area-continuous vector measures, whose value *c(x,n) at an internal body point x* is: (1) tested by driving through x an arbitrary imaginary surface oriented by its normal n, a surface seen as the common boundary of two adjacent body parts; (2) interpreted as the force per unit area about x exerted by the body part laying where n is pointing onto the adjacent body part; (3) represented as the linear action on n of the Cauchy stress field S at x:

$$c(x, n) = S(x)n.$$

This relationship constructs the contact-interaction mapping in terms of the stress field; conversely, as it is well-known, S can be constructed in terms of c as follows:

$$S(x) = \sum_{i=1}^{3} c(x, n_i) \otimes n_i,$$

where n_i are three mutually orthogonal unit vectors, and $\otimes$ denotes the dyadic product of vectors. Moreover, *at a point x of the body's boundary* where the outer normal is $n(x)$ and the applied force per unit area $l(x)$, the contact-interaction mapping enters the stress BC:

$$S(x)n(x) = l(x)$$

in an implicit way, because $S(x)n(x) = c(x, n(x))$. Thus, in standard cases, the spatial regularity of contact interactions, stress fields, and applied loads, is essentially the same.

Now, load concentrations of various types have been considered long since in the classical linear elasticity theory. Needless to say, they induce singularities in the accompanying equilibrium stress fields; the study of these singularities has led to a full understanding of the role of Green kernels, in linear elasticity and in other theories. That *concentrated loads might induce internal concentrated contact interactions* looks like a natural expectation. To incorporate these concentrations, a broad generalization of the standard treatment of contact interactions sketched above is needed; one such generalization has been proposed by Schuricht (2007a, 2007b).

As always, generalizations may be inspired and facilitated by having a certain number of relevant explicit examples at one's disposal, such as those given by Podio-Guidugli (2004, 2006) by revisiting the equilibrium stress fields encountered when studying Flamant's, Cerruti's, and other problems of linear isotropic elasticity where the applied loads are concentrated. For the reader's convenience, we recall here the example that prompted the study of Podio-Guidugli (2004). With the help of Fig. 4.5a, consider the two-dimensional version of Flamant's problem, where the applied load $f = fe_1$ $(f > 0)$ gives rise to the equilibrium stress field

$$S_F(\rho, \vartheta) = -\frac{2}{\pi}\ f\rho^{-1}\cos\vartheta e(\vartheta) \otimes e(\vartheta), \quad (\rho, \vartheta) \in]0, +\infty[\times\left[-\frac{\pi}{2}, +\frac{\pi}{2}\right].$$

As it is easy to verify, when the body part $\mathscr{P}_\rho$ is imagined as isolated from the rest (Fig.4.5b), there are no diffused contact interactions on the straight parts of the boundary and the diffused contact interactions on the curved part, namely,

$$c_F(\rho, \vartheta) = -\frac{2}{\pi}\ f\rho^{-1}\cos\vartheta e(\vartheta), \quad \vartheta \in [0, +\frac{\pi}{2}],$$

cannot balance the applied force $\frac{1}{2}f$; *a force* $\frac{1}{\pi}f\ e_2$ *applied at the origin is needed,* which can be *interpreted as an internal concentrated contact interaction arising at the point O of the vertical part of the boundary of* $\mathscr{P}_\rho$, whatever the choice of $\rho > 0$. Importantly, all of the examples in Podio-Guidugli (2004, 2006) are *independent of material response*; other examples of the same nature are given in Podio-Guidugli and Schuricht (2011) and interpreted in the light of the generalized theory of contact interactions developed in Schuricht (2007a, 2007b). The purpose of the paper of Chinchaladze et al. (2011) was to show that internal concentrated contact interactions are at times to be expected within the framework of Vekua's hierarchical theory of prismatic shells. There is exemplified and discussed the phenomenology of those interactions, because the authors of the above paper believe that they should be regarded as a possibility both in continuum mechanics and in structure mechanics, although a treatment of contact interactions general enough to accommodate them is still wanted. The above paper focuses on a class of 3D equilibrium problems for linearly elastic isotropic *cusped* prismatic shell-like bodies, being subject to *concentrated force-and-couple loads* at a cusp point of their boundary. The relevant results from Jaiani (2006, 2008a, 2008b) are recapitulated[3]; it is shown that in a Vekua-type 0−order approximation each of those equilibrium problems reduces to *a problem formulated over a* 2D *flat region* (the so-called 'projection' of the prismatic shell-like region under examination), for *a system of partial differential equations involving thickness-stress components* (that is, stress components integrated along the thickness), which do depend generally on the material moduli and the cusp's geometry. The cases when the projection is a semicircle or a semicircle sector, and hence it is expedient to use cylindrical coordinates, are considered in detail. Finally, there are shown that *at times, internal concentrated contact interactions are needed to maintain equilibrium* by means of three explicit examples, strongly reminiscent, respectively, of those encountered in the study of Flamant's, Cerruti's and Carothers' problems. The results can be summarized as follows. When equilibrium problems for cusped prismatic shells subject to concentrated loads are formulated within the framework of I. Vekua's 0-order approximation, (1) *in both the Flamant-type problem and Cerruti-type problem, where the external concentrated force acts in direction* x_1, *force-like internal concentrated interactions do occur, while they do not occur in the Cerruti-type problem, where the external concentrated force acts in direction*

---

[3] If the thickness of the cusped prismatic shell has the form (4.15), the solutions of Flamant's and Cerruti's (Jaiani 1973, 1974a, 1982, 2006) and Carothers' (Jaiani 2008b) type problems have the forms

$$\sigma_r = r^{-1}\varphi_{k_2}(\psi)\sin^{k_2}\psi,\ \tau_{r\psi} = \tau_{\psi r} = 0,\quad \sigma_\psi = 0,\ Z_r = kr^{-1}\sin^{k_2}\psi,\ Z_\psi = 0,$$

and

$$\sigma_r = \frac{\partial \tau_{r\psi}}{\partial\psi},\ \tau_{r\psi} = \tau_{\psi r} = r^{-2}\widetilde{\varphi}_{k_2}(\psi)\sin^{k_2+1}\psi,\ \partial_\psi = 0,\ Z_r = \frac{\partial Z_\psi}{\partial\psi},\ Z_\psi = \widetilde{k}r^{-2}\sin^{k_2+1}\psi,$$

respectively, $Z_z = \nu(\sigma_r + \sigma_\psi)$; σ_r, σ_ψ, $\tau_{\psi r}$, Z_ψ, Z_r, Z_z, are stress tensor components in cylindrical coordinates $(r, \psi = \frac{\pi}{2} - \theta, z)$ integrated over the thickness;

$$\varphi_{k_2}(\psi) := \begin{cases} \gamma_1 e^{a\psi} + \delta_1 e^{-a\psi} & \text{when } \widetilde{\nu}k_2 > 1,\\ \gamma_2 + \delta_2\psi & \text{when } \widetilde{\nu}k_2 = 1,\\ \gamma_3\cos c\psi + \delta_3\sin c\psi & \text{when } 0 \le \widetilde{\nu}k_2 < 1,\end{cases}$$

$$\widetilde{\varphi}_{k_2}(\psi) := \begin{cases} \widetilde{\gamma}_1[e^{a\psi} + e^{a(\pi-\psi)}] & \text{when } \widetilde{\nu}k_2 > 1,\\ \widetilde{\gamma}_2 & \text{when } \widetilde{\nu}k_2 = 1,\\ \widetilde{\gamma}_3[\cos c\psi + \sin c\psi] & \text{when } 0 \le \widetilde{\nu}k_2 < 1,\end{cases}$$

$$a := \sqrt{(k_2+1)(\widetilde{\nu}k_2 - 1)},\quad c := \sqrt{(k_2+1)(1-\widetilde{\nu}k_2)},\quad \widetilde{\nu} := \nu(1-\nu)^{-1},$$

k, γ_i, δ_i $(i = \overline{1,3})$ and $\widetilde{k}$, $\widetilde{\gamma}_i$, $\widetilde{\delta}_i$ $(i = \overline{1,3})$ are constants depending on applied concentrated forces and moments, correspondingly, and on k_2, $\widetilde{\nu}$. For $k_2 = 0$ along with k=0, $\widetilde{k} = 0$, the above solutions coincide with the classical solutions of Flamant's, Cerruti's, and Carother's problems, respectively.

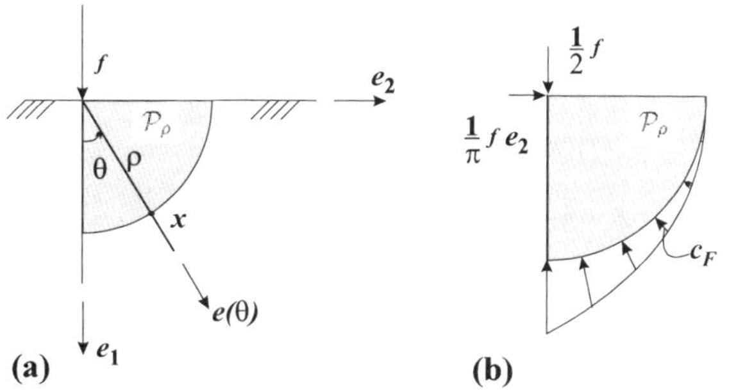

Fig. 4.5 The two-dimensional Flamant's problem: equilibrium of a quarter-disk part

x_3; (2) *in the Carothers-type equilibrium problem, neither force-like nor couple-like internal concentrated interactions occur*. These results parallel closely those obtained, respectively, in Podio-Guidugli (2004, 2006), where the plane versions of Flamant's, Cerruti's, and Carothers', problems were considered.

References

M. Avalishvili, D. Gordeziani, Investigation of two-dimensional models of elastic prismatic shell. Georgian Math. J. **10**(1), 17–36 (2003)

G. Avalishvili, M. Avalishvili, Investigation of a static hierarchic model for elastic shells. Bull. Georgian Acad. Sci. **169**(3), 451–453 (2004)

G. Avalishvili, M. Avalishvili, D. Gordeziani, Construction and investigation of hierarchical models for thermoelastic prismatic shells. Bull. Georgian Acad. Sci. **2**(3), 35–42 (2008)

G. Avalishvili, M. Avalishvili, D. Gordeziani, On some nonclassical two-dimensional models for thermoelastic plates with variable thickness. Bull. Georgian Acad. **4**(2), 27–34 (2010)

K.E. Bisshopp, Lateral bending of symmetrically loaded conical discs. Quart. Appl. Math. **2**(3), 205–217 (1944)

N. Chinchaladze, Vibration of the plate with two cusped edges. Proc. I. Vekua Inst. Appl. Math. **52**, 30–48 (2002d)

N. Chinchaladze, R. Gilbert, G. Jaiani, S. Kharibegashvili, D. Natroshvili, Existence and uniqueness theorems for cusped prismatic shells in the Nth hierarchical model. Math. Methods Appl. Sci. **31**(11), 1345–1367 (2008)

N. Chinchaladze, G. Jaiani, B. Maistrenko, P. Podio-Guidugli, Concentrated contact interactions in cuspidate prismatic shell-like bodies. Archive of Applied Mechanics doi: 10.1007/s00419-010-0496-6, Online First (2011)

M. Dauge, E. Faou, Z. Yosibash, Plates and shells: Asymptotic expansions and hierarchical models in encyclopedia of Computational Mechanics. In: E. Stein, R. Borst, T.J.R. Hugles (eds) *Fundamentals*, vol. 1, (John Wiley & Sons, New York, 1996)

G.G. Devdariani, On the solution with finite energy for bending equation of the prismatic sharpened shell (Russian). Proc. I. Vekua Inst. Appl. Math. **47**, 17–25 (1992)

G. Devdariani, G.V. Jaiani, S.S. Kharibegashvili, D. Natroshvili, The first boundary value problem for the system of cusped prismatic shells in the first approximation. Appl. Math. Inform. **5**(2), 26–46 (2000)

G. Devdariani, The first boundary value problem for a degenerate elliptic system. Bull. TICMI **5**, 23–24 (2001)

G. Fichera, Sulle equazioni defferenziali lineari ellitico-paraboliche del secondo ordine (Italian). Atti Accad. naz. Lincei, Mem. CI. sci. fis., mat. e matur. Sez. I **5**, 1–30 (1956)

G. Fichera, On a unified theory of boundary value problems for elliptic-parabolic equations of second order. Boundary Problems in Differential Equations (Lancer R.R. Ed.), (University of Wisconsin Press, Madison, 1960) pp. 97–120

D.G. Gordeziani, On the solvability of some boundary value problems for a variant of the theory of thin shells (Russian). Dokl. Akad. Nauk SSSR **215**(6), 1289–1292 (1974a)

D.G. Gordeziani, To the exactness of one variant of the theory of thin shells. Soviet. Math. Dokl. **215**(4), 751–754 (1974b)

D. Gordeziani, G. Avalishvili, On approximation of a dynamical problem for elastic mixtures by two-dimensional problems. Bull. Georgian Acad. Sci. **170**(1), 46–49 (2004)

D. Gordeziani, G. Avalishvili, On a dynamical hierarchical model for prismatic shells in the theory of elastic mixtures. Math. Meth. Appl. Sci. **28**, 737–756 (2005)

D. Gordeziani, M. Avalishvili, G. Avalishvili, Dynamical hierarchical models for elastic shells. Appl. Math. Inform. Mech. **10**(1), 19–38 (2005)

D. Gordeziani, M. Avalishvili, G. Avalishvili, Hierarchical models of elastic shells in curvilinear coordinates. Comput. Math. Appl. **51**, 1789–1808 (2006)

G.V. Jaiani, On the deflections of thin wedge-shaped shells (Russian). Bull. Georgian Acad. Sci. **65**(3), 543–546 (1972)

G.V. Jaiani, On one problem of an elastic wedge-shaped body (Russian). Bull. Georgian Acad. Sci. **70**(3), 557–560 (1973)

G.V. Jaiani, On one problem for the fourth-order degenerate equation (Russian). Bull. Georgian Acad. Sci. **75**(1), 17–20 (1974a)

G.V. Jaiani, On a wedge-shaped body arbitrarily loaded along the cusped edge (Russian). Bull. Georgian Acad. Sci. **75**(2), 309–312 (1974b)

G.V. Jaiani, *On a wedge-shaped shell. Proceedings of X All-union Conference in Shell and Plate Theory*. Tbilisi "Metsniereba", (1975) (Russian) pp. 94–101

G.V. Jaiani, Cylindrical bending of a rectangular plate with power law changing stiffness (Russian). *Proceedings of Conference of Young Scientists in Mathematics*. Tbilisi University Press, (1976) pp. 49–52

G.V. Jaiani, On the bending of a plate with power law changing stiffness (Russian). Annuaire Des Ecoles Superieures Mechanique Tech. **12**(2), 15–19 (1977)

G.V. Jaiani, On a physical interpretation of Fichera's function. Acad. Naz. dei Lincei, Rend. della Sc. Fis. Mat. e Nat. S. 8, 68, fasc. **5**, 426–435 (1980a)

G.V. Jaiani, *On some boundary value problems for prismatic cusped shells. Theory of Shells. Proceedings of the Third IUTAM Symposium on Shell Theory Dedicated to the Memory of Academician I.N.Vekua*, ed. by W.T. Koiter, G.K. Mikhailov, North-Holland Publishing Company, (1980b), pp. 339–343

G.V. Jaiani, *Solution of Some Problems for a Degenerate Elliptic Equation of Higher Order and Their Applications to Prismatic Shells* (Russian). (Tbilisi University Press, Tbilisi, 1982)

G.V. Jaiani, *Boundary value problems of mathematical theory of prismatic shells with cusps* (Russian). *Proceedings of All-union Seminar in Theory and Numerical Methods in Shell and Plate Theory*. Tbilisi University Press, (1984a) pp. 126–142

G.V. Jaiani, *Euler-Poisson-Darboux Equation (in Russian, Georgian and English summaries)*. (Tbilisi University Press, Tbilisi, 1984b)

G.V. Jaiani, *Bending of prismatic cusped shell with power law changing stiffness* (Russian). *Proceedings of XIV All-union Conference in Shell and Plate Theory*. Tbilisi University Press, (1987a) pp. 474–479

G.V. Jaiani, *Linear elliptic equations with order degeneration and prismatic shells with cusps* (Russian). *Proceedings of All-union Conference in Modern Problems of Mathematical Physics*. Tbilisi University Press, **1**, 207–213 (1987)

G.V. Jaiani, The first boundary value problem of cusped prismatic shell theory in zero approximation of Vekua's theory (Russian). Proc. I. Vekua Inst. Appl. Math. **29**, 5–38 (1988)

G.V. Jaiani, On a way of construction of solutions of boundary value problems for higher order equations. Rend. Mat. S. **7**(10), 717–731 (1990)

G.V. Jaiani, On some applications of variational methods in the bending theory of prismatic sharpened shells. Proc. I. Vekua Inst. Appl. Math. **47**, 36–44 (1992)

G.V. Jaiani, The main boundary value problems of bending of a cusped plate in weighted sobolev spaces. Appl. Math. Inf. **1**(1), 78–84 (1996)

G. Jaiani, Bending of an orthotropic cusped plate. Appl. Math. Inform. **4**(1), 29–65 (1999a)

G. Jaiani, Oscillation of an orthotropic cusped plate, reports of enlarged session of the seminar of I. Vekua Inst. Appl. Math. Tbilisi State Univ. **14**(1), 24–31 (1999b)

G. Jaiani, *Initial and boundary value problems for singular differential equations and applications to the theory of cusped bars and plates. Complex methods for partial differential equations* (ISAAC Serials, vol. 6), ed. by H. Begehr, O. Celebi, W. Tutschke (Kluwer, Dordrecht, 1999c) pp. 113–149

G.V. Jaiani, Application of Vekua's dimension reduction method to cusped plates and bars. Bull. TICMI **5**, 27–34 (2001)

G.V. Jaiani, Theory of cusped Euler-Bernoulli beams and Kirchoff-Love plates. Lect. Notes TICMI 3 (2002)

G. Jaiani, *Mathematical Models of Mechanics of Continua* (in Georgian). (Tbilisi University Press, Tbilisi, 2004a)

G. Jaiani, Bending of a cusped plate on an elastic foundation. Bull. TICMI **8**, 22–31 (2004b)

G.V. Jaiani, A cusped prismatic shell-like body with the angular projection under the action of a concentrated force. Rendiconti Academia Nazionale delle Scienze detta dei XL, Memorie di Matematica e Applicazioni. 124^0 XXX, facs. 1, 65–82 (2006)

G.V. Jaiani, On a nonlocal boundary value problem for a system of singular differential equations. Appl. Anal. **87**(1), 83–97 (2008a)

G.V. Jaiani, A cusped prismatic shell-like body under the action of concentrated moments. Z. Angew. Math. Phys. **59**, 518–536 (2008b)

G. Jaiani, N. Chinchaladze, Cylindrical bending of a cusped plate with big deflections. J. Math. Sci. **157**(1), 52–69 (2009)

G. Jaiani, A. Kufner, Oscillation of cusped Euler-Bernoulli beams and Kirchhoff-Love plates. Hacettepe J. Math. Stat. **35**(1), 7–53 (2006)

G. Jaiani, B.-W. Schulze, Some degenerate elliptic systems and applications to cusped plates. Mathematische Nachrichten **280**(4), 388–407 (2007)

G. Jaiani, S. Kharibegashvili, D. Natroshvili, W.L. Wendland, Hierarchical models for elastic cusped plates and beams. Lect. Notes TICMI 4 (2003)

G. Jaiani, S. Kharibegashvili, D. Natroshvili, W.L. Wendland, Two dimensional hierarchical models for prismatic shells with thickness vanishing at the boundary. J. Elast. **77**(2), 95–122 (2004)

A.D. Kovalenko, *Circular Plates of Variable Thickness*. (Gos. Izdat of Fis.-Mat. Lit., Moscow, Russian, 1959)

A.R. Khvoles, The general representation for solutions of equilibrium equations of prismatic shell with variable thickness (Russian). Semin. I. Vekua Inst. Appl. Math. Rep. Annot. Rep. **5**, 19–21 (1971)

E.V. Makhover, Bending of a plate of variable thickness with a cusp edge (Russian). Sci. Notes of Leningrad State Ped. Inst. **17**(2), 28–39 (1957)

E.V. Makhover, On the spectrum of the fundamental frequencies of a plate with a cusped edge (Russian). Sci. Notes of Leningrad State Ped. Inst. **197**, 113–118 (1958)

S.G. Mikhlin, *Variational Methods in Mathematical Physics* (Russian). (Nauka, Moscow, 1970)

B. Miara, G. Avalishvili, M. Avalishvili, D. Gordeziani, Hierarchical modeling of thermoelastic plates with variable thickness. Anal. Appl. **8**(2), 125–159 (2010)

P. Podio-Guidugli, Examples of concentrated contact interactions in simple bodies. J. Elast. **75**, 167–186 (2004)

P. Podio-Guidugli, On concentrated contact interactions. Progress in Nonlinear Differential Equations and their Applications **68**, 137–147 (2006)

P. Podio-Guidugli, F. Schuricht, Concentrated actions on cuspidate bodies. J. Elast. (pub. online: 18 Dec 2010) (2011). doi: 10.1007/s10659-010-9295-0

C. Schwab, A-posteriori modelling error estimation for hierarchic plate models. Numer. Math. **74**, 221–259 (1996)

F. Schuricht, Interactions in continuum physics. In: M. Šilhavý (ed.), Mathematical Modeling of Bodies with Complicated Bulk and Boundary Behavior Quaderni di Matematica, **20**, 169–196 (2007a)

F. Schuricht, A new mathematical foundation for contact interactions in continuum physics. Arch. Rat. Mech. Anal. **184**, 495–551 (2007b)

G.V. Tsiskarishvili, N.G. Khomasuridze, Axially symmetric strained state of a cusped conical and cylindrical shells (Georgian). Proc. I. Vekua Inst. Appl. Math. **42**, 59–71 (1991a)

G.V. Tsiskarishvili, N.G. Khomasuridze, Cylindrical bending of a cusped cylindrical shell (Georgian). Proc. I. Vekua Inst. Appl. Math. **42**, 72–79 (1991b)

G.V. Tsiskarishvili, *Elastic Equilibrium of Cusped Plates and Shells* (Georgian). (Tbilisi University Press, Tbilisi, 1993)

Chapter 5
Cusped Beams

Abstract If we consider the cylindrical bending of a plate, in particular, of a cusped one, with rectangular projection $a \leq x_1 \leq b, \quad 0 \leq x_2 \leq L$, then we, actually (if we take Poisson's ratio equal to zero, they will exactly coincide), get the corresponding results also for cusped beams [see (Jaiani, Cylindrical bending of a rectangular plate with power law changing stiffness.Tbilisi University Press, Tbilisi, pp. 49–52, 1976, Boundary value problems of mathematical theory of prismatic shells with cusps. Tbilisi University Press, Tbilisi, pp. 126–142, 1984) , see also (Jaiani, ZAMM Z. Angew. Math. Mech. 81(3):147–173, 2001) and (Chinchaladze, Rep. Enlarged Sess. Semin. I. Vekua Appl. Math 10(1):21–23, 1995 , Rep. Enlarged Sess. Semin. I. Vekua Appl. Math 14(1):12–20, 1999, Proc. I. Vekua Inst. Appl. Math 52:30–48 2002)] . In the present chapter we briefly sketch all the papers devoted to both cusped Euler–Bernoulli beams and hierarchical models of cusped beams.

Keywords Cusped Euler-Bernoulli beams · Hierarchical models of cusped beams · Boundary value problems · Initial boundary value problems

5.1 Cusped Euler–Bernoulli Beams

Kirchhoff (1879) constructed solutions for the harmonic eigenvibration of beams having a form either of a wedge (see Fig. 2.25) or a conical wedge (see Fig. 2.18) by means of Bessel functions. In 1916, using the above exact solution for the wedge shaped beam, Timoshenko (1972) investigated the eigenvibration when cusped beam end is free while the non-cusped end is built-in. In this connection we should also refer to books of Dauge (1988) and Kozlov et al. (1997), where elliptic BVPs in domains with conical or cuspidal boundary points like shown in Figs. 2.18, 2.19 are

G. Jaiani, *Cusped Shell-Like Structures*, SpringerBriefs in Applied Sciences and Technology, DOI: 10.1007/978-3-642-22101-9_5,

considered. The authors concentrate on estimates for solutions in usual and weighted Sobolev spaces of arbitrary integer order, solvability of the BVP, regularity assertions, and, which is very important, asymptotic formulas for the solutions near singular points. This results could be useful by investigation of BVPs for the conical (see Fig. 2.18) and cuspidal (see Fig. 2.19) beams as 3D bodies.

Uzunov (1980) numerically solved the problem of bending of the cusped circular beam (see Fig. 2.17) on an elastic foundation with a constant compliance. The moment of inertia of the beam has the form

$$I(x_2) = \frac{\pi\, r^4}{4}, \quad r = c\, x_2^{\gamma}, \quad c,\ \gamma = const > 0, \quad \gamma < 1$$

(r is the cross-section radius). The blunt cusped end is free and the non-cusped one is clamped.

In 1990–1995 the bending vibration of homogeneous Euler–Bernoulli cone beams and beams of continuously varying rectangular cross-sections, when one side (width) of the cross-section is constant, while the other side (thickness) has the form (4.15), were considered in Naguleswaran (1990, 1992, 1994, 1995); first, the concrete cases of $k_2 = 1,\ 1/2$, and finally, the general case of k_2 were studied. In these investigations the cusped end is always free; direct analytical solutions were constructed for the mode shape equation and the frequencies were also tabulated.

Two contact problems were considered in Shavlakadze (1999, 2001), [see also (2007)]. Namely, the contact problem is studied for an unbounded elastic medium composed of two welded half-planes $x_1 > 0$ and $x_1 < 0$ having different elastic constants and strengthened on the semi-axis $x_2 > 0$ by an inclusion of variable thickness (a cusped thin beam) with constant Young's modulus and Poisson's ratio. It is assumed that the plate is subjected to plane strain, the flexural rigidity D of the inclusion has the form

$$D = D_0 x_2^{\alpha}, \quad D_0,\ \alpha = const > 0,$$

and the cusped end $x_2 = 0$ of the beam is free. A case when the cusped inclusion is orthogonal to the axes along which the half-planes are welded is considered in Shavlakadze (2010). The second contact problem considered is the problem of bending (under action of bending moments $M_{x_2}^{\infty} = M,\ M_{x_1}^{\infty} = 0$ at infinity) of an isotropic plate of a constant thickness reinforced by a finite elastic rib (beam) with the flexural rigidity D of the form

$$D = (a^2 - x_2^2)^{n+\frac{1}{2}} P(x_2),$$

where $a = const > 0,\ n \geq 1$ is an integer and $P(x_2)$ is a polynomial which may have only simple roots and satisfies certain additional restrictions. It is assumed that the rib is not loaded and the rib ends are free. Presence of the inclusion involves a jump of the generalized transverse force

$$N_{x_2}(0-, x_2) - N_{x_2}(0+, x_2) = \mu(x_2)$$

with the unknown contact force $\mu(x_2)$ of interaction between the inclusion and the plate.

By Jaiani (2002a, 2002b) [see also Jaiani and Kufner (2006)] for cusped Euler–Bernoulli beams all the reasonable BVPs are solved in explicit (integral) forms. Moreover, the existence and uniqueness of weak (generalized) solutions to vibration problems in suitably chosen weighted Sobolev spaces are established. The governing equation under consideration has the form

$$\left[E(x_2)I(x_2)w_{,22}\right]_{,22} = q(x_2,t) - \rho\sigma(x_2)\frac{\partial^2 w}{\partial t^2}, \quad 0<x_2<L, \tag{5.1}$$

where $w(x_2,t)$ is a deflection of the beam, $q(x_2,t)$ is an intensity of the load, $E(x_2)$ is the Young modulus, $\rho(x_2)$ is a density, $\sigma(x_2)$ is the area of the cross-section, $I(x_2)$ is the moment of inertia with respect to the barycentric axis normal to the plane x_2x_3. The product $E(x_2)I(x_2)$ may vanish at the beam ends $x_2 = 0$ and $x_2 = L$. Posing of BCs depends on convergence-divergence of the integrals

$$I_\alpha^0 := \int_0^\varepsilon \tau^\alpha E^{-1}(\tau)I^{-1}(\tau)d\tau, \quad I_\alpha^L := \int_{L-\varepsilon}^L (L-\tau)^\alpha E^{-1}(\tau)I^{-1}(\tau)d\tau, \quad \alpha = 1,2,$$

in a one-sided ε neighborhoods of the beam ends $x_2 = 0$ and $x_2 = L$.

On the cusped edge $x_2 = 0$ (correspondingly, $x_2 = L$) we can admit the following classical BCs:

$$w = w_0 \text{ (correspondingly, } w_L), \quad w_{,2} = w_0' \text{ (correspondingly, } w_L') \tag{5.2}$$
$$\text{iff } I_0^0 \text{ (correspondingly, } I_0^L) < +\infty;$$

$$w_{,2} = w_0' \, (w_L'), \quad Q_2 = Q_0(Q_L) \text{ iff } I_0^0 \, (I_0^L) < +\infty; \tag{5.3}$$

$$w = w_0 \, (w_L), \quad M_2 = M_0(M_L) \neq 0 \text{ iff } I_1^0 \, (I_1^L) < +\infty; \tag{5.4}$$

$$M_2 = M_0(M_L), \quad Q_2 = Q_0(Q_L) \text{ if } I_0^0 \, (I_0^L) \leq +\infty, \tag{5.5}$$

and the following nonclassical (in the sense of the bending theory) conditions (replacing BCs):

$$w = w_0 \, (w_L), \quad w_{,2} = O(1) \text{ when } x_2 \to 0+ \; (x_2 \to L-) \tag{5.6}$$

if

$$I_0^0 \, (I_0^L) = +\infty, \quad I_1^0 \, (I_1^L) < +\infty;$$

$$w = O(1), \quad w_{,2} = O(1) \text{ when } x_2 \to 0+ \; (x_2 \to L-) \quad \text{if } I_1^0 \, (I_1^L) = +\infty, \tag{5.7}$$

where $w_0, w_L, w_0', w_L', M_0, M_L, Q_0, Q_L$ are given constants, O is a Landau symbol ($O(1)$ means boundedness).

Theorem 5.1 Let $f \in C(]0,L[)$, $D \in C^2(]0,L[) \cap C([0,L])$. *Then the following BVPs are uniquely solvable and stable, i.e., well-posed in the sense of Hadamard:*

1.(5.1), $(5.2)_0$ $(5.2)_L$, $w \in C^4(]0,L[) \cap C^1([0,L])$;
2.(5.1), $(5.3)_0$ $(5.2)_L$, $w \in C^4(]0,L[) \cap C^1([0,L])$;
3.(5.1), $(5.4)_0$ $(5.2)_L$, $w \in C^4(]0,L[) \cap C^1(]0,L]) \cap C([0,L])$;
4.(5.1), $(5.5)_0$ $(5.2)_L$, $w \in C^4(]0,L[) \cap C^1(]0,L])$;
5.(5.1), $(5.2)_0$ $(5.3)_L$, $w \in C^4(]0,L[) \cap C^1([0,L])$;
6.(5.1), $(5.4)_0$ $(5.3)_L$, $w \in C^4(]0,L[) \cap C^1(]0,L]) \cap C([0,L])$;
7.(5.1), $(5.2)_0$ $(5.4)_L$, $w \in C^4(]0,L[) \cap C^1([0,L[) \cap C([0,L])$;
8.(5.1), $(5.3)_0$ $(5.4)_L$, $w \in C^4(]0,L[) \cap C^1([0,L[) \cap C([0,L])$;
9.(5.1), $(5.4)_0$ $(5.4)_L$, $w \in C^4(]0,L[) \cap C([0,L])$;
10.(5.1), $(5.2)_0$ $(5.5)_L$, $w \in C^4(]0,L[) \cap C^1([0,L[)$;
11.(5.1), $(5.7)_0$, $(5.2)_L$, $w \in C^4(]0,L[) \cap C^1(]0,L])$;
12.(5.1), $(5.2)_0$, $(5.7)_L$, $w \in C^4(]0,L[) \cap C^1([0,L[)$;
13.(5.1), $(5.6)_0$, $(5.2)_L$, $w \in C^4(]0,L[) \cap C^1(]0,L]) \cap C([0,L])$;
14.(5.1), $(5.2)_0$, $(5.6)_L$, $w \in C^4(]0,L[) \cap C^1([0,L[) \cap C([0,L])$;
15.(5.1), $(5.6)_0$, $(5.6)_L$, $w \in C^4(]0,L[) \cap C([0,L])$.

Indices 0 and L at (5.2)–(5.7) *mean the corresponding formulas for the points 0 and L, respectively.*

5.2 (N_3, N_2) Approximation

The well-possedness of BVPs in weighted displacements for (0,0) and (1,0) approximations are investigated in [Jaiani (1998a, 1998b),respectively]. Setting of BCs in the general (N_3, N_2) approximation is considered in Jaiani (2001). In the (N_3, N_2) approximation if at an end of the bar $2h_i > 0$, $i = 2, 3$, then the BCs on this end have the forms as follows.

1. BCs in displacements:

$$v_{jrs} = f_{jrs}, \quad j = \overline{1,3}, \quad r = \overline{0, N_3}, \quad s = \overline{0, N_2}, \tag{5.8}$$

   where constants f_{jrs} are weighted double moments of displacements given at the end.

2. BCs in stresses:

$$X_{j1rs} = g_{jrs}, \quad j = \overline{1,3}, \quad r = \overline{0,N_3}, \quad s = \overline{0,N_2}, \tag{5.9}$$

   where constants g_{jrs} are double moments of the stress vector components given at the end.
3. Mixed BCs are called such BCs when either at one end conditions (5.8), while at the other end conditions (5.9), or at both ends for some indices conditions (5.8) and for the remaining ones conditions (5.9) are given.

In the dynamical case we have to add the initial conditions as well:

$$v_{jrs}|_{t=0} = \varphi_{jrs}(x_1), \quad \left.\frac{\partial v_{jrs}}{\partial t}\right|_{t=0} = \psi_{jrs}(x_1), \tag{5.10}$$

$$x_1 \in]0, L[, \;\; j = \overline{1,3}, \quad r = \overline{0,N_3}, \quad s = \overline{0,N_2},$$

where functions φ_{jrs}, ψ_{jrs} are weighted double moments of displacements and velocities given, and f_{jrs} and g_{jrs} should be, in general, functions of t.

If at an end of a bar at least one of $2h_i$, $i = 2, 3$, vanishes then in conditions (5.8), (5.9) the left-hand side terms should be understood as limits from the inside of the bar, provided that these limits exist. In some cases conditions (5.8) should be replaced by conditions of boundedness of v_{jrs}.

To investigate the above problems one can use the vast literature on ordinary differential equations [see e.g. Kiguradze (1987, 1997)], and on hyperbolic partial differential equations [see e.g. Ladyzhenskaya (1973)]. But to apply them, one needs special efforts especially in the case $2h_i(x_1) \geq 0$, $i = 2, 3$, on $[0, L]$.

If in the static case $h_i > 0$, $i = 2, 3$, on $[0, L]$, then the existence of a regular solution in the classical sense, taking into account that the system (3.15) can be reduced to the system of first order ordinary differential equations, by virtue of the well-known theorem [see e.g. Kiguradze (1997), p. 146], follows from the uniqueness of the solution of the above problems which can be proved by means of the potential energy analogously to Vekua (1955).

If in the static case $2h_i|_{x_1=0,L} \geq 0$ and $2h_i(x_1) > 0$, $i = 2, 3$, on $]0, L[$, the following BCs can be posed in the (N_3, N_2) approximation [see Jaiani (2001), Sect. 5]:

$$v_{jn_3n_2}(0) = \varphi^0_{jn_3n_2}, \;\; j = \overline{1,3}, \quad \text{if} \;\; \overset{n_3,n_2}{I_0} < +\infty, \;\; n_i = \overline{0,N_i}, \;\; i = 2, 3; \tag{5.11}$$

$$v_{jn_3n_2}(L) = \varphi^L_{jn_3n_2}, \;\; j = \overline{1,3}, \quad \text{if} \;\; \overset{n_3,n_2}{I_L} < +\infty, \;\; n_i = \overline{0,N_i}, \;\; i = 2, 3; \tag{5.12}$$

$$v_{jn_3n_2}(x_1) = O(1), \; x_1 \to 0+, \; j = \overline{1,3}, \;\; \text{if} \;\; \overset{n_3,n_2}{I_0} = +\infty, \; n_i = \overline{0,N_i}, \quad i = 2, 3; \tag{5.13}$$

$$v_{jn_3n_2}(x_1) = O(1),\ x_1 \to L-,\ j=\overline{1,3},\ \text{if}\ \overset{n_3,n_2}{I_L} = +\infty,\ n_i = \overline{0,N_i},\quad i=2,3; \tag{5.14}$$

$$\begin{aligned}
&\lim_{x_1\to 0+} h_2^{n_2} h_3^{n_3} X_{1jn_3n_2}(x_1)\\
&= \lim_{x_1\to 0+} \Bigg\{ \Lambda_j h_2^{2n_2+1} h_3^{2n_3+1}\, v_{jn_3n_2,1} - \Lambda_j \sum_{k=2}^{3} \sum_{s=n_k+1}^{N_k} h_2^{\delta_{k2}s+(\delta_{k3}+1)n_2+1}\\
&\times h_3^{\delta_{k3}s+(\delta_{k2}+1)n_3+1} h_k^{-1} \overset{n_k}{b_s^k}\, v_{j\,\delta_{k2}n_3+\delta_{k3}s\,\delta_{k2}s+\delta_{k3}n_2} - \left[\delta_{j1}\lambda\sum_{i=2}^{3} + \delta_{ij}(\delta_{j2}+\delta_{j3})\mu\right]\\
&\times \sum_{s=n_i+1}^{N_i} h_2^{\delta_{i2}s+(\delta_{i3}+1)n_2+1} h_3^{(\delta_{i2}+1)n_3+\delta_{i3}s+1} h_i^{-1} \overset{n_i}{b_{is}} v_{\delta_{j1}i+\delta_{j2}+\delta_{j3}\ \delta_{i2}n_3+\delta_{i3}s\,\delta_{i2}s+\delta_{i3}n_2} \Bigg\}\\
&= \psi^0_{jn_3n_2},\ j=\overline{1,3},\ \text{if}\ \overset{n_3,n_2}{I_0} \le +\infty,\ n_i=\overline{0,N_i},\quad i=2,3;
\end{aligned} \tag{5.15}$$

$$\begin{aligned}
&\lim_{x_1\to L-} h_2^{n_2} h_3^{n_3} X_{1jn_3n_2}(x_1) = \psi^L_{jn_3n_2} \quad j=\overline{1,3},\\
&\text{if}\ \overset{n_3,n_2}{I_L} \le +\infty,\ n_i=\overline{0,N_i},\quad i=2,3;
\end{aligned} \tag{5.16}$$

where $\varphi^0_{jn_3n_2}$, $\varphi^L_{jn_3n_2}$, $\psi^0_{jn_3n_2}$, $\psi^L_{jn_3n_2}$ are given constants,

$$\overset{n_3,n_2}{I_0} := \int_0^{\varepsilon} h_2^{-2n_2-1}(\tau) h_3^{-2n_3-1}(\tau)d\tau,\quad \varepsilon = const > 0,$$

$$\overset{n_3,n_2}{I_L} := \int_{L-\varepsilon}^{L} h_2^{-2n_2-1}(\tau) h_3^{-2n_3-1}(\tau)d\tau,\quad \varepsilon = const > 0.$$

The BVPs $(3.15)_0$,[1] (5.11), (5.12); $(3.15)_0$, (5.11), (5.16); $(3.15)_0$, (5.12), (5.15); $(3.15)_0$, (5.11), (5.14); $(3.15)_0$, (5.12), (5.13) cannot have more than one solution (here we do not take care to make precise the appropriate classes of solutions). The problem (3.15), (5.13), (5.14) can have a solution up to the rigid-body motion [see Jaiani (2001), Sect. 8].

If in a neighborhood of a cusped beam end 3D stresses are bounded, then at this end all moments of the stress vector will be equal to zero. Non-zero stress vector moments given at a cusped beam end mean that in 3D model this end is loaded by a concentrated at the cusped endpoint or cusped edge surface force.

---

[1] Under $(3.15)_0$ we mean the static system corresponding to (3.15). In what follows we use similar stipulation in other cases as well.

In the dynamical case we have to add to the BCs (5.11)–(5.16) the initial conditions (5.10) since $h_i(x_1) > 0$, $i = 2, 3$, for $t = 0$, $x_1 \in]0, L[$, and, therefore, the system (3.15) is not degenerate for such (x_1, t).

In the $(0,0)$ approximation the system (3.15) will have the form

$$(h_2 h_3 v_{j,1}(x_1, t))_{,1} + \overset{0,0}{Y_j} = \Lambda_j^{-1} \rho h_2 h_3 \frac{\partial^2 v_j(x_1, t)}{\partial t^2}, \; j = \overline{1, 3},$$
$$v_j(x_1, t) := \frac{u_{j00}(x_1, t)}{h_2(x_1) h_3(x_1)}, \quad \overset{0,0}{Y_1} := \frac{\overset{0,0}{X_1^0}}{\lambda + 2\mu}, \quad \overset{0,0}{Y_i} := \frac{\overset{0,0}{X_1^0}}{\mu}, \quad i = 2, 3. \tag{5.17}$$

Let us consider the static case of (5.17). Obviously, we can represent the general solution of $(5.17)_0$ as follows

$$v_j(x_1) = -\int\limits_{x_1^0}^{x_1} \frac{d\tau}{h_2(\tau) h_3(\tau)} \int\limits_{x_1^0}^{\tau} \overset{0,0}{Y_j}(t) dt + c_1^j \int\limits_{x_1^0}^{x_1} \frac{d\tau}{h_2(\tau) h_3(\tau)} + c_2^j, \tag{5.18}$$

$$x_1^0 = const \in]0, L[, \quad c_\alpha^j = const, \quad \alpha = 1, 2, \quad j = \overline{1, 3}.$$

Let further

$$v_j(0) = \varphi_j^0 \quad \text{if} \quad \overset{0,0}{I_0} < +\infty, \quad j = \overline{1, 3}, \tag{5.19}$$

$$v_j(L) = \varphi_j^L \quad \text{if} \quad \overset{0,0}{I_L} < +\infty, \; j = \overline{1, 3}, \tag{5.20}$$

$$v_j(x_1) = O(1), \quad x_1 \to 0+, \quad \text{if} \; \overset{0,0}{I_0} = +\infty, \; j = \overline{1, 3}, \tag{5.21}$$

$$v_j(x_1) = O(1), \quad x_1 \to L-, \quad \text{if} \; \overset{0,0}{I_L} = +\infty, \; j = \overline{1, 3}, \tag{5.22}$$

$$X_{1j00}(0) = \Lambda_j \lim_{x_1 \to 0+} h_2 h_3 v_{j,1} = \psi_j^0 \quad \text{if} \; \overset{0,0}{I_0} \le +\infty, \quad j = \overline{1, 3}, \tag{5.23}$$

$$X_{1j00}(L) = \Lambda_j \lim_{x_1 \to L-} h_2 h_3 v_{j,1} = \psi_j^L \quad \text{if} \; \overset{0,0}{I_L} \le +\infty, \quad j = \overline{1, 3}, \tag{5.24}$$

φ_j^0, ψ_j^0, φ_j^L, ψ_j^L, $j = \overline{1, 3}$, are given constants.

Equations (5.19), (5.20) are called the Dirichlet conditions; (5.23), (5.24) and (5.21), (5.22) are called Neumann and Keldysh conditions, respectively.

Let $\overset{0,0}{Y_j} \in L([0, L])$, and if $\overset{0,0}{I_0} = +\infty$ ($\overset{0,0}{I_L} = +\infty$), let it be such that the iterated integral in (5.18) be bounded as $x_1 \to 0+$ $(L-)$.

If $\frac{1}{h_2 \cdot h_3}$ is locally summable[2] on $]0,L[$, then from (5.18) it follows that for regular solutions ($v_j \in C^2(]0,L[)$) only the following problems are well-posed: $(5.17)_0$, (5.19), (5.20) ($v_j \in C([0,L])$); $(5.17)_0$, (5.21), (5.20) ($v_j \in C(]0,L])$); $(5.17)_0$, (5.19), (5.22) ($v_j \in C([0,L[)$); $(5.17)_0$, (5.21), (5.22) (v_j is bounded); $(5.17)_0$, (5.19), (5.24) ($v_j \in C([0,L[)$, $h_2h_3v_{j,1} \in C(]0,L])$; $(5.17)_0$, (5.23), (5.20) ($v_j \in C(]0,L])$, $h_2h_3v_{j,1} \in C([0,L[)$; the mixed problems when on the one end of the beam for some displacement components the Dirichlet and for others either the Neumann or the Keldysh conditions are given, while on the other end for the first components either the Neumann, or the Dirichlet, or Keldysh conditions and for the second components the Dirichlet conditions (the Neumann and Keldysh conditions are not admissible) are given, are well-posed.

In the case $(5.17)_0$, (5.21), (5.22) we have got the solution up to the rigid-body translation ($c_1^j = 0$, c_2^j, $j = 1,2,3$, are arbitrary). It should be so, since in this case the beam under consideration as a 3D body is loaded only by surface forces (which are included in $\overset{0,0}{Y_j}$) acting on the lateral surfaces of the beam. All other BVPs have unique solutions and constants c_α^j, $\alpha = 1,2$, $j = \overline{1,3}$, can explicitly be calculated.

Remark 5.1 Let

$$2h_i(x_1) = h_0^i x_1^{\kappa_i}(L - x_1)^{\delta_i}, \quad h_0^i = const > 0, \quad \kappa_i, \ \delta_i = const \geq 0, \quad i = 2,3.$$

Then we can give a transparent geometric interpretation of posing of above BVPs. To $I_0(I_L) < +\infty$ there corresponds $\kappa_2 + \kappa_3 < 1$ ($\delta_2 + \delta_3 < 1$), and to $I_0(I_L) = +\infty$ there corresponds $\kappa_2 + \kappa_3 \geq 1$ ($\delta_2 + \delta_3 \geq 1$). If at least one of the beam profile and projection has a sharp cusp (see the definition in Sect. 4.1 and also Sect. 2.1), i.e., $\kappa_2 + \kappa_3 \geq 1$ ($\delta_2 + \delta_3 \geq 1$), then at this end the components of the displacement vector cannot be prescribed. If both the profile and projection have blunt cusps (see the definition in Sect. 4.1 and also Sect. 2.1) and $\kappa_2 + \kappa_3 \geq 1$ ($\delta_2 + \delta_3 \geq 1$), then the same is true but for $\kappa_2 + \kappa_3 < 1$ ($\delta_2 + \delta_3 < 1$) the above components can be given.

If

$$2h_i := h_i^0 x_1^{\alpha_i}, \quad h_i^0, \alpha_i = const > 0, \ i = 2,3, \tag{5.25}$$

equations (5.17) are of the type of equations considered in Kharibegashvili and Jaiani (2001) and according to results of this paper, the edge $x_1 = 0$ of the beam can be fixed only if $\alpha = \alpha_2 + \alpha_3 < 1$. Moreover, there exist unique strong generalized solutions of two IBVPs: (5.17), (5.25), (5.10) when $N_2 = N_3 = 0$, (5.19), (5.20) and (5.17), (5.25), (5.10) when $N_2 = N_3 = 0$, (5.21), (5.20). The dynamical problems in (1,0) approximation are studied as well. In [Dobrushkin (1988),

[2] i.e., integrable on any interval contained in $]0,L[$

Chap. 5] in the case $\alpha_3 = 0$, $\alpha_2 = 1$, the BVP, when the prescribed on the sides of the wedge displacements u_r, u_ψ stabilize at the spike to constants and vanish at infinity, is investigated within the framework of the classical plane deformation.

Kharibegashvili and Jaiani (2000) have studied the well-possedness of dynamical problems in (0,0) approximation with mixed BCs at beam ends when either integrated stresses or weighted displacement components are prescribed. In definite cases the last conditions should be replaced by Keldysh conditions.

By Chinchaladze et al. (2010) a dynamical problem in the (0,0) approximation of elastic cusped prismatic beams is investigated when stresses are applied at the lateral surfaces and the ends of the beam. Two types of cusped ends are considered when the beam cross-section turns into either a point or a straight line segment. Correspondingly, at the cusped end either a force concentrated at the point or forces concentrated along the straight line segment are applied. The existence and uniqueness theorems in appropriate weighted Sobolev spaces are proved.

Some problems for cusped beams when they occupy the Lipschitz 3D domains are investigated by Gordeziani and Avalishvili (2003); Avalishvili (2004, 2006, 2009).

References

G. Avalishvili, M. Avalishvili, On a hierarchical model of elastic rods with variable cross-sections. Appl. Math. Inform. Mech. **9**(1), 1–16 (2004)

G. Avalishvili, M. Avalishvili, Investigation of dynamical one-dimensional models for elastic rods with variable cross-sections. Bull. Georgian. Acad. Sci. **174**(3), 399–402 (2006)

G. Avalishvili, M. Avalishvili, On the investigation of one-dimensional models for thermoelastic beams. Bull. Georgian Acad. **3**(3), 25–32 (2009)

N. Chinchaladze, Cylindrical bending of the prismatic shell with two sharp edges in case of a strip. Rep. Enlarged Sess.Semin. I. Vekua Appl. Math. **10**(1), 21–23 (1995)

N. Chinchaladze, On the vibration of an elastic cusped plate. Rep. Enlarged Sess. Semin. I. Vekua Appl. Math. **14**(1), 12–20 (1999)

N. Chinchaladze, Vibration of the plate with two cusped edges. Proc. I. Vekua Inst.Appl. Math. **52**, 30–48 (2002)

N. Chinchaladze, R. Gilbert, G. Jaiani, S. Kharibegashvili, D. Natroshvili, Cusped elastic beams under the action of stresses and concentrated forces.. Appl. Anal. **89**(5), 757–774 (2010)

V.A. Dobrushkin, *Boundary Value Problems of Dynamical Theory of Elasticity for Wedge-shaped Domains* (Russian). (Nauka i Tekhnika, Minsk, 1988)

M. Dauge, *Elliptic Boundary Value Problems on Corner Domains*. (Springer, Berlin, 1988)

D. Gordeziani, G. Avalishvili, On the investigation of a static hierarchic model for elastic rods. Appl. Math. Inform. Mech. **8**(1), 34–46 (2003)

G.V.Jaiani, *Cylindrical bending of a rectangular plate with power law changing stiffness* (Russian). *Proceedings of Conference of Young Scientists in Mathematics*. (Tbilisi University Press, Tbilisi, 1976), pp. 49–52

G.V.Jaiani, *Boundary value problems of mathematical theory of prismatic shells with cusps* (Russian). *Proceedings of All-union Seminar in Theory and Numerical Methods in Shell and Plate Theory*. (Tbilisi University Press, Tbilisi, 1984), pp. 126–142

G. Jaiani, On a model of a bar with variable thickness. Bull. TICMI **2**, 36–40 (1998a)

G. Jaiani, *On a Mathematical Model of a Bar with Variable Rectangular Cross-section*.Preprint 98/21. (Institut fuer Mathematik, Universitaet Potsdam, 1998b)

G. Jaiani, On a mathematical model of bars with variable rectangular cross-sections. ZAMM Z. Angew. Math. Mech. **81**(3), 147–173 (2001)

G.V. Jaiani, Theory of cusped Euler–Bernoulli beams and Kirchoff-Love plates. Lect. Notes TICMI 3 (2002a)

G.V. Jaiani, Static and dynamical problems for a cusped beam. Proc. I. Vekua Inst. Appl. Math. **52**, 1–29 (2002b)

G. Jaiani, A. Kufner, Oscillation of cusped Euler-Bernoulli beams and Kirchhoff-Love plates. Hacettepe J. Math. Stat. 35 (1), 7–53 (2006)

S.S. Kharibegashvili, G. Jaiani, Dynamical problems in the (0,0) and (1,0) approximations of a mathematical model of bars. In "Functional-analytic and complex methods, interactions, and applications to PDEs". *Proceedings of the International Graz Workshop*, Graz, Austria, 12–16 Feb 2001, World Scientific, pp. 188–248

S. Kharibegashvili, G. Jaiani, On a vibration of an elastic cusped bar. Bull. TICMI. **4**, 24–28 (2000)

G. Kirchhoff, Über die Transversalschwingungen eines Stabes von veränderlichen Querschnitt. Monatsbericht der Königlich Preussischen Akademie der Wissenschaften zu Berlin. Sitzung der physikalisch-mathematischen Klasse Oktober, (1879), pp. 815–828

V.A. Kozlov, V.G. Maz'ya, J. Rossmann, *Elliptic Boundary Value Problems in Domains with Point Singularities*. (American Mathematical Society, Providence RI, 1997)

I.T. Kiguradze, Initial and boundary value problems for the systems of ordinary differential equations (Russian). Itogi Nauki i Tekhniki, Seria Sovremennje Problemj Mathematiki, Noveyshie Dostizhenia V. 30, Moskva 1–202 (1987)

I.T. Kiguradze, *Initial and boundary value problems for the systems of ordinary differential equations (Russian). V. 1. Linear Theory*. (Metsniereba, Tbilisi, 1997)

O.A. Ladyzhenskaya, *Boundary Value Problems of Mathematical Physics* (Russian). (Nauka Press, Moscow, 1973)

S. Nagulcswaran, The vibration of an Euler-Bernoulli beam of constant depth and with convex parabolic variation in breadth. Nat. Conf. Publ. Inst. Eng. Austr. **9**, 204–209 (1990)

S. Naguleswaran, Vibration of an Euler-Bernoulli beam of constant depth and with linearly varying breadth. J. Sound Vib. **153**(3), 509–522 (1992)

S. Naguleswaran, The direct solution for the transverse vibration of Euler-Bernoulli wedge and cone beams. J. Sound Vib. **172**(3), 289–304 (1994)

S. Naguleswaran, The vibration of a "complete" Euler-Bernoulli beam of constant depth and breadth proportional to axial coordinate raised to a positive exponent. J. Sound Vib. **187**(2), 311–327 (1995)

N. Shavlakadze, Contact problem of the interaction of a semi-finite inclusion with a plate. Georgian Math. J. **6**(5), 489–500 (1999)

N. Shavlakadze, Contact problem of bending of a plate with a thin reinforcement (Russian). Mechanika Tverdogo Tela **3**, 144–150 (2001)

N. Shavlakadze, Contact problems of the mathematical theory of elasticity for plates with an elastic inclusion. Acta Appl. Math. **99**, 29–51 (2007)

N. Shavlakadze, Contact problem for piecewise homogeneous orthotropic plane with finite inclusion. Proceedings of the International Conference "Topical Problems of Continuum Mechanics" **2**, 213–215 (2010)

S.P. Timoshenko, *Course of Theory of Elasticity* (Russian). (Naukova Dumka, Kiev, 1972)

S.G. Uzunov, Variational-difference Approximation of a Degenerate Ordinary Differential Equation of Fourth Order (Russian). in Collection: Correct Boundary Value Problems for Non-classical Equations of Mathematical Physics, Novosibirsk, (1980), pp. 159–164

I.N. Vekua, On one method of calculating of prismatic shells (Russian). Trudy Tbilis. Mat. Inst. **21**, 191–259 (1955)

Chapter 6
Relations of 3D, 2D, and 1D Problems

Abstract In this chapter papers concerning relations of 3D, 2D, and 1D problems for cusped bodies are surveyed. Main attention is paid to peculiarities of posing BCs for cusped bodies and to their relation to 3D BCs. This matter is thoroughly discussed in both the cases of BCs in stresses and in displacements. Significance and relations of physical and mathematical moments of stresses are emphasized.

Keywords Mathematical moments • Physical moments • Nonclassical boundary conditions in elasticity

6.1 Boundary Conditions in Stresses

The work of Jaiani (2008) deals with the analysis of the physical and geometrical sense of Nth ($N = 0, 1, \ldots$) order moments and weighted moments of the stress tensor and the displacement vector in the theory of cusped prismatic shells. The peculiarities of the setting of BCs at cusped edges in terms of moments and weighted moments are analyzed. The relation of such BCs to the BCs of the 3D theory of elasticity is also discussed. Now, more in detail. The kth order mathematical moment of $X_{ij} \in C(\overline{\Omega}\backslash\overline{\Gamma})$ is defined by (3.1).

We define physical moments [compare with the mathematical moments (3.1)] as follows:

$$M_{ijk}(x_1,x_2) := \int\limits_{\overset{(-)}{h}(x_1,x_2)}^{\overset{(+)}{h}(x_1,x_2)} X_{ij}(x_1,x_2,x_3)x_3^k dx_3, \quad k = 0,1,\ldots, \quad i,j = 1,2,3, \quad (6.1)$$

G. Jaiani, *Cusped Shell-Like Structures*, SpringerBriefs in Applied Sciences and Technology, DOI: 10.1007/978-3-642-22101-9_6,

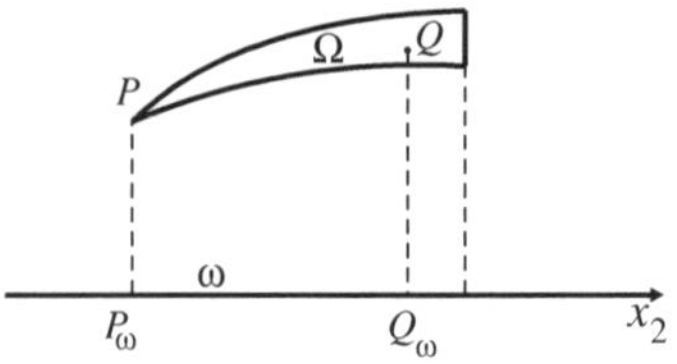

Fig. 6.1 A prismatic shell profile (*cross-section*), when $P \neq P_\omega$

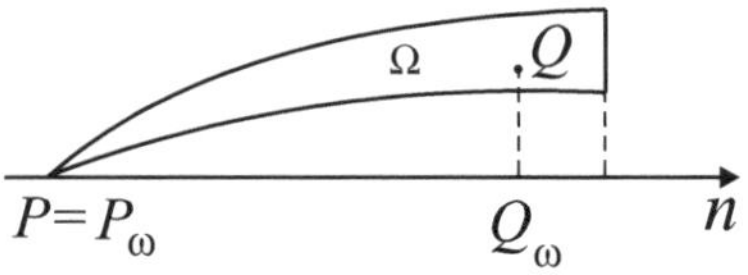

Fig. 6.2 A prismatic shell profile, when $P = P_\omega$ (nonsymmetric case)

$$S_{ij}(x_1,x_2) := M_{ij0}(x_1,x_2), \quad i,j = 1,2,3,$$

where S_{23}, S_{13} are the so-called shear forces, $S_{\alpha\beta}$, $\alpha,\beta = 1,2$ are the so-called membrane (or normal and tangent) forces, M_{111}, M_{221} are the bending moments and M_{121} is the twisting moment. In what follows, generalizing these definitions, $M_{ijk}(x_1,x_2)$ will be called physical moments of the kth order.

It is clear that (6.1) and (3.1) make sense for $2h > 0$.

If a point P belongs to the cusped edge $\Gamma_0 \subseteq \Gamma$ of the shell, i.e., if

$$2h(P_\omega) = 0, \;\; P_\omega \in \gamma_0,$$

(see Fig. 6.1), then the mathematical and physical moments we define as the following limits:

$$X_{ijk}(P_\omega) := \lim_{\omega \ni Q_\omega \to P_\omega} X_{ijk}(Q_\omega), \;\; i,j = \overline{1,3},$$

$$M_{ijk}(P_\omega) := \lim_{\omega \ni Q_\omega \to P_\omega} M_{ijk}(Q_\omega), \;\; i,j = \overline{1,3}, \tag{6.2}$$

where $\gamma_0 \subseteq \partial\omega$, P_ω and Q_ω are the projections of $P \in \Gamma$ and $Q \in \Omega$, respectively. Clearly, Γ_0 is a curve lying on the cylindrical surface bounding the shell from the side (see Fig. 6.4 and also Figs. 2.9, 2.10, 2.12). It is evident that γ_0 is the projection of Γ_0. In what follows, by the normals n at points of Γ_0, normals at the same points to the above cylindrical surface are meant. When the cusped edge lies on $\partial\omega$, then obviously $P_\omega \equiv P$ (see Figs. 6.2, 6.3, where normal sections of the prismatic shell are shown).

When functions X_{ij} are bounded on Ω and $P \in \Gamma_0$, then $X_{ijk}(P_\omega) = 0$ and $M_{ijk}(P_\omega) = 0$.

$$X_{ijk}(P_\omega) \neq 0 \;\text{ and }\; M_{ijk}(P_\omega) \neq 0 \;\text{ only if }\; \lim_{\Omega \ni Q \to P \in \Gamma_0} X_{ij}(Q) = \infty.$$

Suppose that some neighborhoods of the point $P \in \Gamma_0$ along Γ_0 and on the upper and lower face surfaces are not loaded. Forces and physical moments

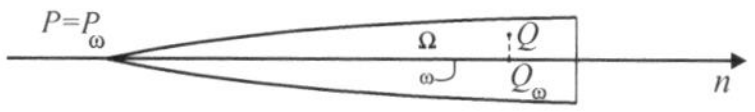

Fig. 6.3 A prismatic shell profile (*cross-section*), when $P = P_\omega$ (symmetric case)

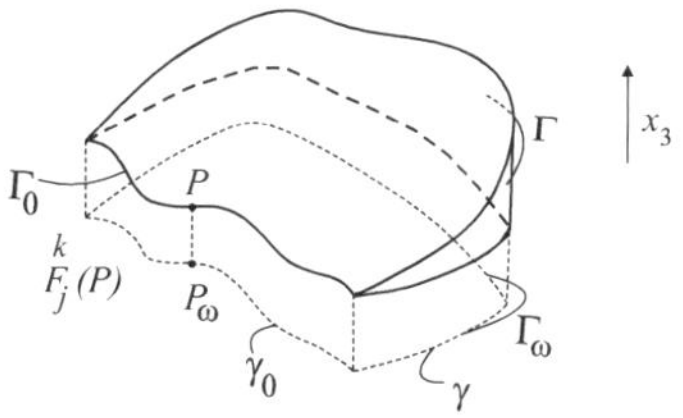

Fig. 6.4 A force concentrated at a point P of a cusped edge

concentrated at the point P of the cusped edge of the shell are defined as follows (see Fig. 6.4):

$$\overset{k}{F}_j(P) := \lim_{\rho\to 0}\int_S X_{nj}(Q)x_3^k dS = \lim_{\rho_\omega\to 0}\int_S X_{nj}(Q_\omega,x_3)x_3^k dS$$

$$= \lim_{\rho_\omega\to 0}\int_{S_\omega} dS_\omega \int_{\overset{(-)}{h}(x_1,x_2)}^{\overset{(+)}{h}(x_1,x_2)} X_{nj}(Q_\omega,x_3)x_3^k dx_3 = \lim_{Q_\omega\to P_\omega}\int_{S_\omega} M_{njk}(Q_\omega)dS_\omega,\ \ j=\overline{1,3},$$

where S is a lying inside the prismatic shell arbitrary cylindrical surface with the generatrix parallel to $0x_3$, S_ω is its projection on the plane $x_3 = 0$; ρ is maximum distance between P and $Q \in S$, and ρ_ω is the maximum distance between P_ω and $Q_\omega \in S_\omega$.

Suppose that there are no forces and moments concentrated at any point. Forces and physical moments concentrated along an arc $d\Gamma_0$ (the upper and lower face surfaces are not loaded) are defined as follows:

$$\overset{k}{E}_{nj}(P)d\Gamma_0 := \lim_{d\to 0}\int_{S_0\cup\Delta'\cup\Delta''} X_{nj}(Q)x_3^k dS$$

$$= \lim_{d\to 0}\int_{\overset{(-)}{h}(x_1,x_2)}^{\overset{(+)}{h}(x_1,x_2)} X_{nj}(Q_\omega,x_3)x_3^k dx_3 dl_0 + \lim_{d\to 0}\int_0^d dl' \int_{\overset{(-)}{h}(x_1,x_2)}^{\overset{(+)}{h}(x_1,x_2)} X_{nj}(Q_\omega,x_3)x_3^k dx_3$$

$$+ \lim_{d\to 0}\int_0^d dl'' \int_{\overset{(-)}{h}(x_1,x_2)}^{\overset{(+)}{h}(x_1,x_2)} X_{nj}(Q_\omega,x_3)x_3^k dx_3$$

$$= \lim_{Q_\omega\to P_\omega} M_{njk}(Q_\omega)dl_0 = M_{njk}(P_\omega)d\Gamma_0 = M_{ijk}(P_\omega)n_i(P_\omega)d\Gamma_0,$$

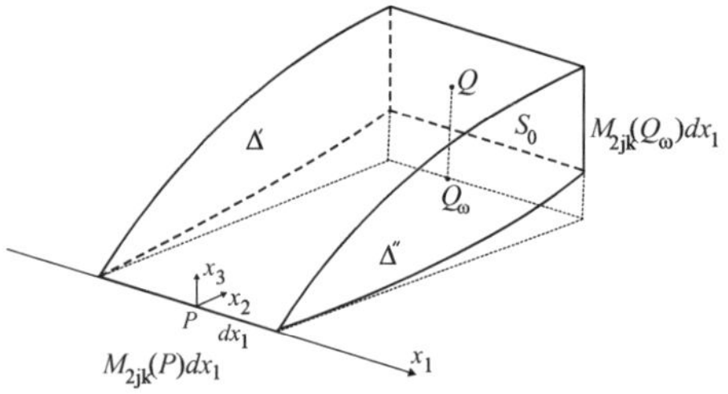

Fig. 6.5 Moments concentrated along a cusped edge

where S_0 is the part of a cylindrical surface parallel to the lateral cylindrical surface passing through the point P, cut out by the upper and lower face surfaces and the planes (their parts containing in the body we denote by Δ' and Δ'') passing through the endpoints of $d\Gamma_0$ orthogonal to the lateral cylindrical surface where Γ_0 lies, d is the distance between the above cylindrical surfaces, dl_0 is the arc lying on S_0 parallel to $d\Gamma_0$ (for the case $P \equiv P_\omega$, $d\Gamma_0 \equiv dx_1$, $dl_0 \equiv dx_1$, see Fig. 6.5). $\overset{k}{E}_{nj}(P)$, $j = \overline{1,3}$, are the components of a line-concentrated forces ($k = 0$) and moments ($k \geq 1$) at the point P. In the case $P \equiv P_\omega$ we have $\overset{k}{E}_{nj}(P) = M_{njk}(P)$ [compare with (6.2)].

Remark 6.1 Since the body is in equilibrium, it follows that

$$\int\limits_{S_0 \cup \Delta' \cup \Delta''} X_{nj}(Q) x_3^k dS = \int\limits_{S_{d\Gamma_0}} X_{nj}(Q) x_3^k dS_{d\Gamma_0},$$

for all arbitrary cylindrical surfaces $S_{d\Gamma_0}$ lying inside the body parallel to the axis Ox_3 and passing through the endpoints of $d\Gamma_0$.

By virtue of (6.1) and

$$P_k(\tau) := \sum_{l=0}^{\left[\frac{k}{2}\right]} (-1)^l \frac{(2k-2l)!}{2^k l!(k-l)!(k-2l)!} \tau^{k-2l}, \quad k = 0, 1, \ldots, \tag{6.3}$$

where $\left[\frac{k}{2}\right]$ is the integer part of $\frac{k}{2}$ [see Whittaker and Watson (1927), p. 302], we have [see Jaiani (2008)],

$$M_{ijk}(Q_\omega) = \int\limits_{\overset{(-)}{h}}^{\overset{(+)}{h}} x_3^k X_{ij}(Q_\omega, x_3) dx_3 = \frac{2^k (k!)^2}{(2k)!} h^k X_{ijk}(Q_\omega) - \frac{(k!)^2}{(2k)!} \left\{ \sum_{l=1}^{\left[\frac{k}{2}\right]} \sum_{r=0}^{k-2l} + \sum_{\substack{r=1 \\ l=0}}^{k} \right\}$$
$$(-1)^{l+r} \frac{(2k-2l)!}{l!(k-l)!r!(k-2l-r)!} \tilde{h}^r h^{2l} M_{ijk-2l-r}(Q_\omega),$$
$$\sum_{l=1}^{0} (\cdot) \equiv 0, k = 0, 1, 2, \cdots. \tag{6.4}$$

This result yields the recurrence formulas that allow to calculate X_{ijk} by means of M_{ijs}, $s = \overline{0,k}$, and M_{ijk} by means of X_{ijs}, $s = \overline{0,k}$.

For instance, for $k = 0, 1$,

$$M_{ij0}(Q_\omega) = X_{ij0}(Q_\omega), \tag{6.5}$$

$$M_{ij1}(Q_\omega) = hX_{ij1}(Q_\omega) + \tilde{h}X_{ij0}(Q_\omega), \tag{6.6}$$

$$hX_{ij1}(Q_\omega) = M_{ij1}(Q_\omega) - \tilde{h}M_{ij0}(Q_\omega). \tag{6.7}$$

Now, letting Q tend to P, i.e., Q_ω to P_ω, from (6.4) we obtain (see Fig. 6.1):

$$\begin{aligned}\overset{k}{E}_{ij}(P) := M_{ijk}(P_\omega) = \frac{2^k (k!)^2}{(2k)!} \lim_{\omega \ni Q_\omega \to P_\omega} h^k(Q_\omega) X_{ijk}(Q_\omega) \\ - k! \sum_{r=1}^{k} (-1)^r \frac{1}{(k-r)!r!} \tilde{h}^r(P_\omega) M_{ijk-r}(P_\omega),\end{aligned} \tag{6.8}$$

since $l > 0$ in the double sum of (6.4) and $h(P_\omega) = 0$.

Note that $\tilde{h}(P_\omega)$ equals the x_3-coordinate of the point P. Hence, when $P \equiv P_\omega$, we have $\tilde{h}(P_\omega) = 0$ and, therefore, from (6.8) we get:

$$M_{ijk}(P_\omega) = \frac{2^k (k!)^2}{(2k)!} \lim_{\omega \ni Q_\omega \to P} h^k(Q_\omega) X_{ijk}(Q_\omega).$$

When we consider BCs in terms of stresses, we find natural to prescribe as BCs the physical moments

$$M_{njk}(P_\omega) = M_{ijk}(P_\omega) n_i(P_\omega) \tag{6.9}$$

concentrated along the cusped edge. We remind that the zero moments are concentrated forces [see (6.5)]. Then in the Nth approximation for the mathematical moments at the cusped edge from (6.8), (6.9) [see also (6.5)–(6.7)] we get the following BCs:

$$\begin{aligned}&\lim_{\omega \ni Q_\omega \to P_\omega} h^k(Q_\omega) X_{njk}(Q_\omega) \\ &= \frac{(2k)!}{2^k k!} \left[\frac{1}{k!} M_{njk}(P_\omega) + \sum_{r=1}^{k} (-1)^r \frac{1}{(k-r)!r!} \tilde{h}^r(p_\omega) \times M_{njk-r}(P_\omega) \right], \\ &\sum_{r=1}^{0}(\cdot) \equiv 0, \; j = \overline{1,3}, \text{ are prescribed for } k = \overline{0,N},\end{aligned} \tag{6.10}$$

which are weighted BCs for $k \geq 1$.

The homogeneous BCs (6.10) at cusped edges of the 2D model correspond to the 3D model, when on the face surfaces stress vectors and on the lateral non-cusped edge (boundary) $\Gamma \backslash \overline{\Gamma}_0$ either the displacements or the stress vectors are prescribed. In this instance, the homogenous BCs (6.10) are automatically satisfied for the bounded stresses or for $u_i \in H^1(\Omega)$, since in the last case $X_{ij} \in L_2(\Omega)$ and applying the Fubini theorem the summability of X_{ij} along x_3 in (3.1) can be proved. Therefore,

$$\lim_{\omega \ni Q_\omega \to P_\omega} \overset{N}{X}_{njk}(Q_\omega) = \lim_{\omega \ni Q_\omega \to P_\omega} \overset{N}{X}_{ijk}(Q_\omega) n_i(Q_\omega) = 0,$$

since the integration limits in (3.1) tend to 0. So, the homogeneous BCs (6.10) at cusped edges are not real BCs; BCs are replaced by the requirement that the desired quantities belong to certain function spaces defined on ω, where we are looking for solutions without any BCs at the cusped edge. The last case arises in Jaiani et al. (2004).

The nonhomogeneous BCs (6.10) at cusped edges of the 2D model correspond to the 3D model, when at the above-mentioned cusped edge Γ_0 forces and physical moments concentrated along the cusped edge are applied, while on the other parts $\overset{(+)}{h}$, $\overset{(-)}{h}$, $\Gamma \backslash \overline{\Gamma}_0$ of the body boundary the same conditions as in the above-formulated case of the homogeneous BCs (6.10) are given.

In the Nth approximation the stress tensor is represented in the form:

$$X_{ij}(Q) \cong \sum_{k=0}^{N} a\left(k+\frac{1}{2}\right) \overset{N}{X}_{ijk}(Q_\omega) P_k(ax_3 - b), \quad i,j = \overline{1,3}. \tag{6.11}$$

We remind (see Sect. 3.4) that the relation (3.1) between X_{ij} and $\overset{N}{X}_{ijk}$ is correct when the symbol of approximate equality in (6.11) is replaced by the exact equality symbol. This occurs if either X_{ij} are Nth order polynomials with respect to x_3 or $N = +\infty \Big\{$ i.e., we consider the Fourier–Legendre series representation (see Sect. 3.4) for

$$X_{ij}(Q_\omega, \cdot) \in C^2\left(\left[\overset{(-)}{h}(Q_\omega), \overset{(+)}{h}(Q_\omega)\right]\right)\Big\}.$$

Surface forces $X_{ni}(P)$ can be considered only at points of blunt cusped edges (in this case the union of the upper and lower surfaces is a smooth surface and there exist normals to the above surface at points $P \in \Gamma_0$). As it follows from (6.11), (6.10) $X_{ni}(Q)$ become infinite as $Q \to P$ if the BCs (6.10) are inhomogeneous. For example, for $N = 0$, it follows from (6.11) that

$$X_{ij}(Q) = \frac{1}{2h} \overset{0}{X}_{ij0}(Q_\omega), \quad \overset{0}{X}_{ij0}(P_\omega) \neq 0.$$

The last means that along the cusped edges forces concentrated along the edge are given [see Figs. 6.6, 6.7, where the plane sections of the corresponding 3D problems are given in the case of symmetric prismatic shells (i.e., plates)]. Some

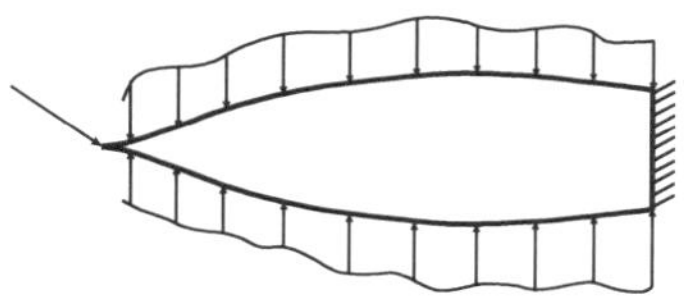

Fig. 6.6 A concentrated force applied at a sharp cusped edge

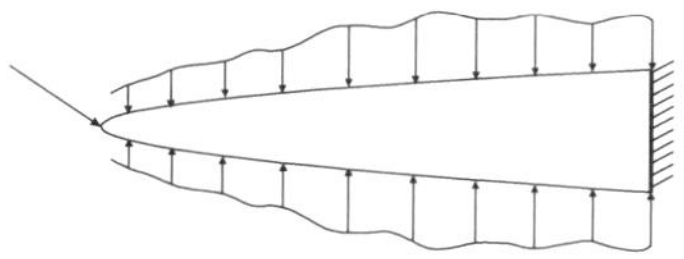

Fig. 6.7 A concentrated force applied at a blunt cusped edge

such problems in 2D and 1D formulations are solved by Jaiani (1980a, 1982, 2001a). A similar analysis can be carried out for cusped beams.

6.2 Boundary Conditions in Displacements

In Jaiani (2007, 2008) general analysis of setting of correct (with Dirichlet and Keldysh type BCs) BVPs in displacements in the Nth ($N = 0, 1, \ldots$) approximation of hierarchical models of cusped prismatic shells (Vekua models) and beams (Jaiani models) is carried out. The relation of the above problems to the corresponding 3D problems is also treated [see also Jaiani (2002)]. In Jaiani (2007) the special attention is paid to the models when on the face surfaces of the prismatic shells and beams displacements are prescribed. The kth order moments of the displacement vector components are defined by (3.1). If u_i, $i = \overline{1,3}$, are bounded functions on Ω, then for $P \in \Gamma_0$, evidently, $u_{ik}(P_\omega) = 0$.

In the Nth approximation, in view of (3.5), (6.3), u_i are expressed by u_{ik} as follows:

$$
\begin{aligned}
u_i(x_1,x_2,x_3) &\cong \sum_{k=0}^{N} a\left(k+\frac{1}{2}\right) u_{ik}(x_1,x_2)P_k(ax_3-b)\\
&= \sum_{k=0}^{N}\left(k+\frac{1}{2}\right) h^k v_{ik}(x_1,x_2)P_k(ax_3-b)\\
&= \sum_{k=0}^{N}\frac{1}{2^k}\left(k+\frac{1}{2}\right)\sum_{l=0}^{\left[\frac{k}{2}\right]}(-1)^l\frac{(2k-2l)!}{l!(k-l)!(k-2l)!}v_{ik}(x_1,x_2)\left(x_3-\tilde{h}\right)^{k-2l}h^{2l},\\
& i=\overline{1,3}, \qquad (6.12)
\end{aligned}
$$

where

$$
v_{ik}(x_1,x_2) = \frac{u_{ik}(x_1,x_2)}{h^{k+1}}, \quad i=\overline{1,3}, \quad k=\overline{0,N},
$$

are weighted moments of the displacements (WMDs).

In particular, in the $N=0$ and $N=1$ approximations respectively

$$u_i(Q) \cong \frac{1}{2} v_{i0}(Q_\omega), \text{ and } u_i(Q) \cong \frac{1}{2} v_{i0}(Q_\omega) + \frac{3}{2} v_{i1}(Q_\omega)\big(x_3 - \tilde{h}\big), \; i = \overline{1,3}. \tag{6.13}$$

Below in the corresponding approximations "$\cong$" is replaced by "=".

Let v_{ik} be bounded, then for

$$k - 2l + 2l = k > 0,$$

since on the one hand, when $P \neq P_\omega$, then $x_3 \to x_3^0 \neq 0$ as $Q \to P$ (when $P \equiv P_\omega$, evidently $x_3^0 = 0$) and on the other hand $\lim\limits_{Q_\omega \to P_\omega} \tilde{h}(Q_\omega) = x_3^0$ (see Fig. 6.1), the limits on the right-hand side of (6.12) are zero as $\Omega \ni Q \to P$, i.e., as $\omega \ni Q_\omega \to P_\omega$. Hence, there remains only the summand for $k = 0$, i.e.,

$$u_i(P) := \lim_{\Omega \ni Q \to P} u_i(Q) = \frac{1}{2} v_{i0}(P_\omega) \text{ if } I_0 := \int\limits_{P_\omega}^{Q_\omega} \frac{dn}{h} < +\infty, \tag{6.14}$$

where n is an inward normal to $\partial\omega$ at the point P_ω, since for $I_0 = +\infty$ the limit

$$\lim_{\omega \ni Q_\omega \to P_\omega} v_{i0}(Q_\omega),$$

in general, can not be assigned [see Jaiani (1996, 2001b, 2007) and references therein].

From (6.12), evidently,

$$\frac{\partial^j u_i(Q)}{\partial x_3^j} = \sum_{k=j}^{N} \frac{1}{2^k}\left(k + \frac{1}{2}\right) \sum_{l=0}^{\left[\frac{k-j}{2}\right]} (-1)^l \frac{(2k-2l)!}{l!(k-l)!(k-2l)!} v_{ik}(Q_\omega)(k-2l)$$
$$\times (k-2l-1)\cdots(k-2l-j+1)\big(x_3 - \tilde{h}\big)^{k-2l-j} h^{2l}, \quad j = \overline{1,N}, \quad i = \overline{1,3}. \tag{6.15}$$

From (6.15), if additionally v_{ik} are bounded, we get

$$\lim_{\Omega \ni Q \to P} \frac{\partial^j u_i(Q)}{\partial x_3^j} = \frac{1}{2^j}\left(j + \frac{1}{2}\right) \frac{(2j)!}{j!} v_{ij}(P_\omega)$$
$$\text{for all } j = \overline{1,N}, \; i = \overline{1,3}, \text{ when } I_N := \int\limits_{P_\omega}^{Q_\omega} \frac{dn}{h^{2N+1}} < +\infty. \tag{6.16}$$

For a fixed j the last inequality should be replaced by the condition

$$I_j := \int\limits_{P_\omega}^{Q_\omega} \frac{dn}{h^{2j+1}} < +\infty, \tag{6.17}$$

since for $I_j = +\infty$ the limit $\lim\limits_{\omega \ni Q_\omega \to P_\omega} v_{ij}(Q_\omega)$, in general, can not be assigned [see Jaiani (2001b), Jaiani and Schulze (2007), Devdariani et al. (2000), Devdariani (2001), Chinchaladze et al. (2008)].

Thus, from (6.14), (6.16) we can define $v_{ij},\ j=\overline{0,N},\ i=\overline{1,3}$, provided that the left-hand sides of (6.14), (6.16), i.e., displacements $u_i,\ i=\overline{1,3}$, and their derivatives with respect to x_3 up to the Nth order of the displacements u_i at point $P \in \Gamma_0$ are known.

Remark 6.2 (6.14) and (6.16) with (6.17) signify that BCs of 2D models, when on γ_0

$$v_{ij}(P_\omega), \quad j=\overline{0,N}, \quad i=\overline{1,3}, \quad \text{are prescribed,}$$

correspond to the BCs of the 3D model, when on Γ_0

$$\frac{\partial^j u_i(P)}{\partial x_3^j}, \quad j=\overline{0,N}, \quad i=\overline{1,3}, \quad \text{are prescribed if} \quad \int\limits_{P_\omega}^{Q_\omega} \frac{dn}{h^{2j+1}} < +\infty.$$

Evidently, all the $v_{ij}(P),\ j=\overline{0,N},\ i=\overline{1,3}$, can be prescribed on γ_0 at the same time only if

$$\int\limits_{P_\omega}^{Q_\omega} \frac{dn}{h^{2N+1}} < +\infty, \tag{6.18}$$

which correspond to the BCs of the 3D model, when, under same condition (6.18), on Γ_0

$$\frac{\partial^j u_i(P)}{\partial x_3^j}, \ j=\overline{0,N}, \ i=\overline{1,3}, \quad \text{are prescribed.} \tag{6.19}$$

Note that when on the lateral surface, which is not degenerated into the cusped edge, the 3D displacements are prescribed, i.e.,

$$u_i(x_1,x_2,x_3)\big|_{(x_1,x_2)\in\partial\omega}$$

are given, we can calculate their derivatives provided that the data are sufficiently smooth with respect to x_3 (which will be equivalent to calculation of the mathematical moments of $u_i(x_1,x_2,x_3)\big|_{(x_1,x_2)\in\partial\omega}$). It means that in this case on the lateral boundary displacements with their derivatives with respect to x_3 are prescribed. We could assume them to be prescribed on cusped edges and try to satisfy (6.19) as limits of the derivatives at the body boundary from its inner part, but such BCs will fall outside the limits of the classical 3D theory of elasticity. When $N=+\infty$ (i.e., we actually have 3D model), then

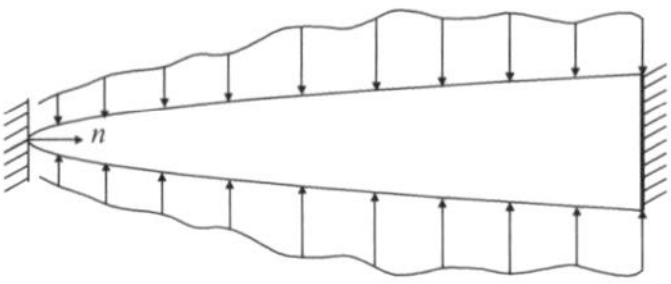

Fig. 6.8 A fixed cusped edge

$$\lim_{N\to+\infty} \int_{P_\omega}^{Q_\omega} \frac{dn}{h^{2N+1}} = +\infty \tag{6.20}$$

and therefore, the above BCs disappear like in the classical 3D model [e.g. if $P_\omega \equiv (x_1, 0)$ and $2h = h_0 x_2^\kappa$, $h_0, \kappa = \text{const} > 0$, then

$$\int_{P_\omega}^{Q_\omega} \frac{dn}{h^{2N+1}} = \frac{1}{h_0^{2N+1}} \int_0^\varepsilon \frac{dx_2}{x_2^{\kappa(2N+1)}} < +\infty,$$

when

$$\kappa(2N+1) < 1.$$

Since for $N \to +\infty$ the last is violated, (6.20) becomes clear]. So, when on the upper and lower face surfaces and non-cusped edge of the shell surface forces are applied, we arrive at the ordinary BVP in stresses for the classical 3D model.

Remark 6.3 Whenever $v_{i0}(P_\omega) = 0$ along the cusped edge, on the one hand, because of $I_0 < +\infty$ [see (6.14)] this cusped edge is blunt one [see Figs. 2.2, 2.3, 2.4, 2.5, 2.6, 2.7, 2.12, Statement 4.1, and Jaiani (1980b)], on the other hand, it means that in the corresponding 3D problem the cusped edge is fixed; let on the face surfaces the stress vectors and on the lateral non-cusped edge either the displacements or the stress vectors be prescribed (see Fig. 6.8). The physical, more precisely geometrical, sense of $v_{i0}(P_\omega) \neq 0$ is evident. Such a formulation of the 3D BVP is not usual and falls outside the limits of the classical 3D theory of elasticity, but the corresponding 2D problem is correct within the framework of the $N = 0$ approximation.

Remark 6.4 If we consider reasonable (strongly speaking, correct) BVPs in corresponding weighted Sobolev spaces, then generalizing known results (see Sect. 4.4) for the Nth approximation at a cusped edge we will get the following BCs in WMDs in the sense of traces (Jaiani 2001b):

$$v_{ij}(P_\omega) \text{ if } \int_{P_\omega}^{Q_\omega} \frac{dn}{h^{2j+1}} < +\infty, \quad j = \overline{0, N}, \quad i = \overline{1, 3}, \quad \text{are given.}$$

Whence, by virtue of (6.12), for displacements we obtain that there exist traces

$$u_i(P),\quad i=\overline{1,3},\quad \text{if}\quad \int\limits_{P_\omega}^{Q_\omega} \frac{dn}{h^{2N+1}} < +\infty.$$

Remark 6.5 When we consider BVPs in displacements, independent of boundary data for 3D body, moments of all the orders of the displacement vector are equal to zero along the cusped edge, provided that the 3D displacement vector is bounded. If we are looking for 2D solutions in 2D weighted Sobolev Spaces which correspond to the H^1 solution of the 3D BVP, then the traces of the above moments are zero along the cusped edge too. It differs from the case of non-cusped edges, where the above moments (we are looking for) depend on the 3D displacement vector values on the non-cusped edge and are not zero, in general. Therefore, the zero values at the cusped edges of the displacement vector unknown moments of the arbitrary order cannot be considered as BCs since we are not able to prescribe their values at the prismatic shell projection boundary arbitrarily. It will be observed that displacement vector components' moments have not geometrical meaning in contrast to WMDs which after multiplication by the corresponding Legendre polynomials in a certain combination give the displacement vector corresponding to the definite hierarchical model [see e.g. (6.13)]. In accordance with the above mentioned fact there arises natural desire to fulfill BCs in the displacements by means of WMDs.

Remark 6.6 Note that in the versions of hierarchical models under consideration the constructed approximate solutions of the corresponding 3D BVPs approximately satisfy BCs at the non-cusped edge and do not satisfy BCs on the face surfaces unless either $N=\infty$ or the solutions of the 3D problem are polynomials with respect to x_3 of the order $\leq N$. Therefore, the same is true for the cusped edge, since the last represents an intersection of the closures of the upper and lower face surfaces. In the cases when the well-posedness of 2D BVPs requires setting of some BCs at the cusped edge, this is caused by the own (inner) nature of the 2D model, which can be considered as an independent model, and not of the nature of the 3D original model.

References

N. Chinchaladze, R. Gilbert,G. Jaiani, S. Kharibegashvili, D. Natroshvili, Existence and uniqueness theorems for cusped prismatic shells in the N-th hierarchical model. Math. Methods in Appl. Sci. **31**(11), 1345–1367 (2008)

G. Devdariani, The first boundary value problem for a degenerate elliptic system. Bull. TICMI **5**, 23–24 (2001)

G. Devdariani, G.V. Jaiani, S.S. Kharibegashvili, D. Natroshvili, The first boundary value problem for the system of cusped prismatic shells in the first approximation. Appl. Math. Inform. **5**(2), 26–46 (2000)

G.V. Jaiani, *On some boundary value problems for prismatic cusped shells. Theory of shells. Proceedings of the Third IUTAM Symposium on Shell Theory Dedicated to the Memory of*

Academician I.N. Vekua, ed. by W.T. Koiter, G.K. Mikhailov (North-Holland Publishing Company, Amsterdam, 1980a), pp. 339–343

G.V. Jaiani, On a physical interpretation of Fichera's function. Acad. Naz. dei Lincei, Rend. della Sc. Fis. Mat. e Nat. S. 8, 68, fasc. **5**, 426–435 (1980b)

G.V. Jaiani, *Solution of Some Problems for a Degenerate Elliptic Equation of Higher Order and Their Applications to Prismatic Shells* (Russian). (Tbilisi University Press, Tbilisi, 1982)

G.V. Jaiani, Elastic bodies with non-smooth boundaries–cusped plates and shells. ZAMM Z. Angew. Math. Mech. **76**(Suppl.2), 117–120 (1996)

G. Jaiani, On a mathematical model of bars with variable rectangular cross-sections. ZAMM Z. Angew. Math. Mech. **81**(3), 147–173 (2001a)

G.V. Jaiani, Application of Vekua's dimension reduction method to cusped plates and bars. Bull. TICMI **5**, 27–34 (2001b)

G.V. Jaiani, Relation of hierarchical models of cusped elastic plates and beams to the three-dimensional models. Semin. I. Vekua Inst. Appl. Math. Rep. **28**, 40–51 (2002)

G.V. Jaiani, *On boundary value problems in displacements for hierachical models of cusped prismatic shells and beams*. Materials of the international conference on non-classic problems of mechanics, vol. 1, Kutaisi, Georgia, 25–27 Oct 2007, pp. 191–196

G. Jaiani, *On physical and mathematical moments and the setting of boundary conditions for cusped prismatic shells and beams. Proceedings of the IUTAM Symposium on Relation of Shell, plate, Beam, and 3D Models, IUTAM Bookseries*, vol. 9 (Springer, New York, 2008), pp. 133–146

G. Jaiani, B.-W. Schulze, Some degenerate elliptic systems and applications to cusped plates. Math. Nachr. **280**(4), 388–407 (2007)

G. Jaiani, S. Kharibegashvili, D. Natroshvili, W.L. Wendland, Two-dimensional hierarchical models for prismatic shells with thickness vanishing at the boundary. J. Elast. **77**(2), 95–122 (2004)

E.T. Whittaker, G.N. Watson, *A Course of Modern Analysis: An Introduction to the General Theory of Infinite Processes and of Analytic Functions; With an Account of the Principal Transcendental Functions*, 4th edn. (Cambridge University Press, Cambridge, 1927), reprinted (1996)

Chapter 7
Cusped Prismatic Shell–Fluid Interaction Problems

Abstract Only a few papers are devoted to cusped elastic structure–fluid interaction problems. In this chapter such works are surveyed when in the solid part either Kirchoff-Love plate or Vekua's zero approximation model and in the fluid part either incompressible ideal or viscous fluids are considered.

Keywords Solid–fluid interaction problems · Kirchoff-Love plate · Vekua's zero approximation · Stoke's equations

Works of Chinchaladze and Jaiani (1998, 2001) and Chinchaladze (2002a, 2002b, 2002c, 2008a) are devoted to some solid–fluid interaction problems when the solid part is an elastic cusped plate. Namely, the adequate transmission conditions on the interface for the interaction problem, when for elastic part the classical Kirchhoff-Love model is used, are studied; the bending of Kirchhoff-Love plates with two cusped edges under action of either incompressible ideal or incompressible viscous fluids has been considered as well, in particular, harmonic vibration is studied. In Chinchaladze and Gilbert (2005, 2006) and Chinchaladze (2008b) an interaction problem of an elastic cusped prismatic shell and an incompressible fluid is investigated by application of the $N = 0$ approximation of Vekua's hierarchical models for the elastic part.

Dynamical 2D models of solid–fluid interaction problems, when solid and fluid parts before modeling occupy 3D Lipschitz domains, are considered in Gordeziani et al. (2009).

In Chinchaladze and Jaiani (2007) two types of hierarchical models for viscous Newtonian fluids occupying thin prismatic domains are constructed (see Sects. 3.6 and 3.7), (i) on the face surfaces only displacements are given; (ii) on the face surfaces only stress vector components are given. In both the cases on lateral cylindrical surfaces arbitrary usual (reasonable) BCs can be prescribed. For hierarchical models for structures consisting of elastic solid and Newtonian fluid parts in prismatic domains interface conditions are established.

In the forthcoming paper of N. Chinchaladze, R. Gilbert, G. Jaiani, S. Kharibegashvili, and D. Natroshvili a mixed dynamical elastic solid–fluid interaction

G. Jaiani, *Cusped Shell-Like Structures*, SpringerBriefs in Applied Sciences and Technology, DOI: 10.1007/978-3-642-22101-9_7,

3D problem is investigated. Applying the Laplace transform technique, the authors reduce the dynamical problem to the elliptic problem which depends on the complex parameter τ and prove the corresponding uniqueness and existence results. Further, the authors establish uniform estimate for solutions and their partial derivatives with respect to the parameter τ at infinity and via the inverse Laplace transform show that the original dynamical transmission problem is uniquely solvable. Further, considering the case when the elastic inclusion is a thin prismatic shell of variable thickness, the authors apply the $N = 0$ approximation of Vekua's hierarchical models for the elastic field in the solid part. This reduces the original problem to the nonclassical BVP for fluid domain with an interior cut along the "middle plane" of the prismatic shell. In contrast to the usual classical streamline conditions, in the case under consideration, on the cut faces there appear non-local BCs. The authors prove unique solvability of the non-classical BVP obtained which leads to the existence results for the original interaction problem with a thin elastic inclusion.

Finally, we find very topical to trace elastic solid–fluid interaction problems when in the ideal and Newtonian fluids cusped beams of type showed in Figs. 2.15, 2.16, 2.17, 2.18, 2.19, 2.20, 2.21, 2.22, 2.23 are submerged.

References

N. Chinchaladze, A cusped elastic plate-ideal incompressible fluid interaction problem. Semin. I. Vekua Inst. Appl. Math. Rep. **28**, 31–39 (2002a)

N. Chinchaladze, On a vibration of an isotropic elastic cusped plates under action of an incompressible viscous fluid. Semin. I. Vekua Inst. Appl. Math. Rep. **28**, 52–60 (2002b)

N. Chinchaladze, On a cusped elastic solid-incompressible fluid interaction problem. Harmonic vibration. Mech. Eng. Tech. Univ. Lodz. **6**(1), 5–29 (2002c)

N. Chinchaladze, On some nonclassical problems for differential equations and their applications to the theory of cusped prismatic shells. Lect. Notes TICMI 9 (2008a)

N. Chinchaladze, *Vibration of an Elastic Plate Under the Action of an Incompressible Fluid. IUTAM Symposium on Relations of Shell, Plate, Beam, and 3D Models, IUTAM Bookser*, vol. 9 (Springer, Dordrecht, 2008b), pp. 77–90

N. Chinchaladze, R. Gilbert, Cylindrical vibration of an elastic cusped plate under the action of an incompressible fluid in case of N = 0 approximation of I. Vekua's hierarchical models. Complex Var. Theor. Appl. **50**(7–11), 479–496 (2005)

N. Chinchaladze, R. Gilbert, Vibration of an elastic plate under action of an incompressible fluid in case of N = 0 approximation of I. Vekua's hierarchical models. Appl. Anal. **85**(9), 1177–1187 (2006)

N. Chinchaladze, G. Jaiani, On a cylindrical bending of a plate with two cusped edges under action of an ideal fluid. Bull. TICMI **2**, 30–34 (1998)

N. Chinchaladze, G. Jaiani, On a cusped elastic solid–fluid interaction problem. Appl. Math. Inform. **6**(2), 25–64 (2001)

N. Chinchaladze, G. Jaiani, *Hierarchical Mathematical Models for Solid–Fluid Interaction Problems* (Georgian). Materials of the International Conference on Non-Classic Problems of Mechanics, Kutaisi, Georgia, 25–27 October, Kutaisi, vol. 2 (2007), pp. 59–64

D. Gordeziani, M. Avalishvili, G. Avalishvili, Dynamical two-dimensional models of solid–fluid interaction. J. Math. Sci. **157**(1), 16–42 (2009)

Index

G. Jaiani, *Cusped Shell-Like Structures*, SpringerBriefs in Applied Sciences and Technology, DOI: 10.1007/978-3-642-22101-9,